Electronics For Dummies®

D1131443

Value Calculations for Ohm's Law

Unknown Value	Formula
Voltage	$V = IR$
Current	$I = V/R$
Power	$P = VI$
Resistance	$R = V/I$

Where,

- ✔ V = voltage (in volts)
- ✔ I = current (in amps)
- ✔ P = power (in watts)
- ✔ R = resistance (in ohms)

Resistor and Capacitor Color Scheme Values

Color	1st Digit	2nd Digit	Multiplier	Tolerance
Black	0	0	x 1	+/- 20%
Brown	1	1	x 10	+/- 1%
Red	2	2	x 100	+/- 2%
Orange	3	3	x 1,000	+/- 3%
Yellow	4	4	x 10,000	+/- 4%
Green	5	5	x 100,000	—
Blue	6	6	x 1,000,000	—
Violet	7	7	x 10,000,000	—
Gray	8	8	x 100,000,000	—
White	9	9	—	—
Gold	—	—	x 0.1	+/- 5%
Silver	—	—	x 0.01	+/- 10%

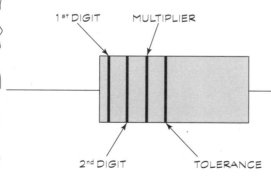

1st DIGIT   MULTIPLIER   2nd DIGIT   TOLERANCE

Note: See Chapter 4 for more information on resistor color codes.

Capacitor Tolerance Markings

Code	Tolerance
B	+/- 0.1 pF
C	+/- 0.25 pF
D	+/- 0.5 pF
F	+/- 1%
G	+/- 2%
J	+/- 5%
K	+/- 10%
M	+/- 20%
Z	+80%, -20%

Capacitor Value Reference

Marking	Value (in micro-farads, μF, unless otherwise noted)
nn	Picofarads (nn is a number from 01 to 99)
101	0.0001
102	0.001
103	0.01
104	0.1
221	0.00022
222	0.0022
223	0.022
224	0.22
331	0.00033
332	0.0033
333	0.033
334	0.33
471	0.00047
472	0.0047
473	0.047
474	0.47

Electronics For Dummies®

Cheat Sheet

Common Component Symbols

Component	Symbol	Component	Symbol
Battery		Phototransistor	
Capacitor		Relay	
Diode		Resistor	
Ground		Speaker	
Interconnections		Switch, SPST	
Photoresistor		Transistor, NPN	

Electronic Component Abbreviations

Term	Abbreviation	Unit	Unit Symbol	Component
Resistance	R	ohm	Ω	Resistor
Capacitance	C	farad	F	Capacitor
Inductance	L	henry	H	Inductor
Voltage	V (or E)	volt	V	
Current	I	amp	A	

Electronic Units of Measure

Number	Name	Scientific Notation	Prefix	Abbreviation
1,000,000,000	1 billion	10^9	giga	G
1,000,000	1 million	10^6	mega	M
1,000	1 thousand	10^3	kilo	k
100	1 hundred	10^2		
10	ten	10^1		
1	1	10^0		
0.1	tenth	10^{-1}		
0.01	hundredth	10^{-2}		
0.001	1 thousandth	10^{-3}	milli	m
0.000001	1 millionth	10^{-6}	micro	μ
0.000000001	1 billionth	10^{-9}	nano	n
0.000000000001	1 trillionth	10^{-12}	pico	p

For Dummies: Bestselling Book Series for Beginners

Electronics FOR DUMMIES®

by Gordon McComb
and Earl Boysen

Wiley Publishing, Inc.

Electronics For Dummies®

Published by
Wiley Publishing, Inc.
111 River Street
Hoboken, NJ 07030-5774

For general information on our other products and services, please contact our Customer Care Department within the U.S. at 800-762-2974, outside the U.S. at 317-572-3993, or fax 317-572-4002.

Wiley also publishes its books in a variety of electronic formats. Some content that appears in print may not be available in electronic books.

Library of Congress Control Number: 2004107905

ISBN: 0-7645-7660-7

Manufactured in the United States of America

10 9 8 7 6 5 4 3 2

1O/RZ/QR/QV/IN

WILEY

About the Authors

Gordon McComb has penned 60 books and over a thousand magazine articles. More than a million copies of his books are in print, in over a dozen languages. For 13 years, Gordon wrote a weekly syndicated newspaper column on personal computers. When not writing about hobby electronics and other fun topics, he serves as a consultant on digital cinema to several notable Hollywood clients.

Earl Boysen is an engineer who, after 20 years in the computer-chip industry, decided to slow down and move to a quiet town in Washington. Earl lives in a house he built with a wonderful lady and finds that he is as busy as ever with teaching, writing, house building, and acting.

Dedication

To my father, Wally McComb, who instilled in me a fascination with electronics; and to Forrest Mims, who taught me a thing or two about it.

G.M.

To my parents, Dick and Nettie, who keep providing an example of the right way to live.

E.B.

Authors' Acknowledgments

The authors give heartfelt thanks to Wiley and the hard-working editors at Wiley, especially Katie Feltman, Nancy Stevenson, Carol Sheehan, Laura Miller, and Amanda Foxworth. Many thanks also to Ward Silver, for his excellent and thorough technical review, and Matt Wagner at Waterside Productions for always having a positive outlook. Author Gordon wishes to thank his family, who once again put their lives on hold while he finished another book.

Publisher's Acknowledgments

We're proud of this book; please send us your comments through our online registration form located at www.dummies.com/register/.

Some of the people who helped bring this book to market include the following:

Acquisitions, Editorial, and Media Development

Project Editor: Nancy Stevenson

Acquisitions Editor: Katie Feltman

Copy Editor: Laura Miller

Technical Editor: Ward Silver

Editorial Manager: Carol Sheehan

Editorial Assistant: Amanda Foxworth

Cartoons: Rich Tennant (www.the5thwave.com)

Production

Project Coordinator: Maridee Ennis

Layout and Graphics: Lauren Goddard, Denny Hager, Joyce Haughey, Barry Offringa, Melanee Prendergast, Jacque Roth, Erin Zeltner

Proofreaders: Amanda Foxworth, Leeann Harney, Jessica Kramer, Carl Pierce, Carol Sheehan, TECHBOOKS Production Services

Indexer: TECHBOOKS Production Services

Publishing and Editorial for Technology Dummies

  Richard Swadley, Vice President and Executive Group Publisher

  Andy Cummings, Vice President and Publisher

  Mary Bednarek, Executive Acquisitions Director

  Mary C. Corder, Editorial Director

Publishing for Consumer Dummies

  Diane Graves Steele, Vice President and Publisher

  Joyce Pepple, Acquisitions Director

Composition Services

  Gerry Fahey, Vice President of Production Services

  Debbie Stailey, Director of Composition Services

Contents at a Glance

Table of Contents

Part III: Putting It on Paper 121

Chapter 6: Reading a Schematic 123

Chapter 7: Understanding the Basics of Electronics Circuits 141

Introduction

A re you thinking about building your own electronic gizmos? Ever wonder how transistors, capacitors, and other building blocks of electronics work? Do you have an interest in finding out how to solder or make your own circuit boards?

Well, you've come to the right place! *Electronics For Dummies* is the key that opens the fun and exciting door of modern electronics. No dry and boring tome, this; what you hold in your hands is a book that gives you just what you need to know to make and troubleshoot your own electronic gadgets.

Why Buy This Book?

Electronics is a huge — no, make that HUGE — subject. Like any science, it consists of a lot of concepts and all sorts of highly complicated mathematical equations. For any really in-depth study of electronics, you need to spend hours and hours memorizing a lot of facts and figures.

But this book takes a different path. It provides you with just what you need to understand the basics of electronics, get to work building electronic circuits, and even tackle a dozen fun projects that you can build in under an hour each for just a few dollars. This book doesn't pretend to answer all your questions about electronics, but it does give you a good grounding in the essentials and makes this exciting science fun!

Why Electronics?

This is a rhetorical question because you no doubt already know why you have an interest in electronics, or you wouldn't have picked up this book. But we'll take a moment to review the things that make electronics well worth your while.

First off, electronics is fun! You get to build things that beep, whir, flash lights, and even move around the room. You acquire skills so that you can work with neat tools and proudly hold your head up at any gathering of electronics geeks.

And don't forget that electronic products are all around us. They make up a growing part of our lives. Some people are content just accepting these gadgets, gizmos, and widgets, but others want to know how they all work. Obviously, you're in that second group, which is definitely the cooler group out there. The science of electronics has advanced to the point that you can now hold a very powerful computer in the palm of your hand. With that computer, you can build something that controls the lighting in your entire house, a robot that vacuums the living room all on its own, or a sensor system that sounds an alarm if somebody tries to get at your collection of 1950s comic books.

Here's the amazing part: You can make electronic gadgets that do these things for just a couple of bucks! At the same time that the art and science of electronics is rapidly advancing, the price for building a circuit that can do something incredibly nifty is dropping like a stone. Unless you're constructing a time machine, or the world's largest robotic rabbit, the typical home-brewed electronics project costs less than dinner for four at a no-frills restaurant. If you're looking for a cool hobby, electronics is one of the least expensive ones around.

Oh, and did I mention that electronics is fun?

You may also want to consider this possibility: People who know the practical side of electronics — what things are, how they work, and how to put them together — can find some really great jobs on the market right now. If you're interested in a career in electronics, make this book your first step to a fun and rewarding new job.

Also, many other hobbies rely on knowledge of electronics in some way. Maybe you're into model railroading. You can figure out how to build your own automated track switchers. Or perhaps you like racing radio-controlled cars. With an understanding of electronics, you may discover how to improve the performance of your car and beat your best friend in the next race. Knowing more about how electronics stuff works can make your other hobbies more fun.

And, last but not least, electronics is fun. Or maybe I mentioned that already?

Foolish Assumptions

This book assumes that you know diddly about electronics. From the very first chapter, we introduce you to basic concepts that you need to master in order to follow what we say in later chapters. But if you already have a handle on the basics, you can easily jump to a later chapter and dive right in. (If you need to know something really important to keep you safe from something such as electrocution, we provide a cross reference to send you back to the relevant chapter for a refresher course.)

You can also use the most-excellent table of contents at the front of the book and the index that Wiley has thoughtfully provided at the back to quickly find the information that you need.

Safety Is Number 1

Reading about electronics is pretty safe. About the worst that can happen is that your eyes get tired from too many late nights with this book. But actually building electronics projects is another matter. Lurking behind the fun of the electronics hobby are high voltages that can electrocute you, soldering irons that can burn you, and little bits of wire that can fly into your eyes when you snip them off with sharp cutters. Ouch!

Safety is *Numero Uno* in electronics. It's so important, in fact, that we devote an entire chapter of this book (Chapter 2) to it. If you're brand new to electronics, please be sure to read this chapter. Don't skip over it, even if you think you're the safest person on earth. Even if you've dabbled in electronics before, it never hurts to refresh your safety memory. When you follow proper precautions, electronics is a very safe and sane hobby. Be sure to keep it that way!

 Although we try to give you great advice about safety throughout, we can't possibly give you every safety precaution in the world in one book. In addition to reading our advice, use your own common sense, read manufacturer's instructions for parts and tools that you work with, and always stay alert.

How This Book Is Organized

Electronics For Dummies is organized so that you can quickly find, read, and understand the information that you want. It's also organized so that if you have some experience with electronics, you can skip chapters and move on to the parts that interest you.

The chapters in this book are divided into parts that also help you find the information that you're looking for quickly and easily.

Part 1: Getting Started in Electronics

Start with Part I if you're brand-spanking new to electronics. Because this book is designed to get you on the road to electronics as quickly as possible, this part has only two chapters, an overall introduction to electronics concepts and safety information. Please read Chapter 2, "Keeping Humans and Gadgets Safe," even if you decide to skip the introduction to electronics you find in Chapter 1.

Part II: Aisle 5, Component Shack: Stocking Up

If you're just starting out in electronics, you probably need a couple of tools. Read about the most important ones in Chapter 3, "Outfitting Your Electronics Bench."

You can't make a project out of thin air. You need resistors, diodes, capacitors, and other building blocks of today's electronics. Chapters 4 and 5 explain what each of the most important electronics components does and how often you use each in building a circuit.

Part III: Putting It on Paper

If you've ever seen an electronics diagram, you probably thought it looked like Egyptian hieroglyphics. In Chapter 6, we explain all about how to read these diagrams (they're called *schematics*), and you can discover how to follow a schematic to determine the basic functionality of a circuit in Chapter 7.

Part IV: Getting Your Hands Dirty

At this point, you're ready to start building your own electronics projects. The chapters in Part IV tell you how to solder and how to use three of the most important testing tools in electronics — the multimeter, logic probe, and oscilloscope. You don't absolutely need the last two to get started with electronics, so you can come back to Chapter 10 a few months from now if you're just starting out.

Part V: A Plethora of Projects

In Chapters 11 and 12, we demonstrate how to build your own circuits. We cover how to construct temporary circuits on something called a solderless breadboard. Then you discover how to produce permanent circuits using several methods, or by designing and ordering printed circuit boards from a supplier. Chapter 13 introduces you to the exciting universe of *microcontrollers,* electronic circuits that you can program to do any of a million things. And finally, in Chapters 14 and 15, you can play with over a dozen fun (and not too complicated!) projects that you can build yourself.

Part VI: The Part of Tens

This part contains several chapters laid out in top-ten-list format. Here, you explore some optional testing tools that you can add to your electronics bench as you gain more experience; get advice about where to find electronics parts; and finally, study useful electronics formulas that don't require a degree in math.

Icons Used in This Book

We're a graphical society, bombarded with images from blockbuster movies and computer games, and so this book uses little graphic icons to visually point out useful information that you may want to know about.

The Tip icon indicates information that may help save you time, headaches, or money (or all three!). These icons tend to point out tasty morsels that make electronics more enjoyable, so don't just skip 'em!

Uh-oh! Something bad is about to happen — if you don't read the text that follows the Warning icon, that is. Some of these point out cautions to avoid personal injury, and others give you advice on avoiding damage to tools, components, circuits, or your pocketbook.

Think of Remember icons as gentle nudges about important ideas or facts that you really should keep in mind while exploring the electronics world. We also use these icons to note where in the book some subject is originally introduced, so you can flip back to those chapters for a refresher, if you need one.

Part I
Getting Started in Electronics

The 5th Wave   By Rich Tennant

In this part . . .

You say you've always wanted to get into electronics, but didn't know where to start? You've come to the right place!

In the chapters ahead we cover the very basics of electrons and electronics: what they're all about and why you should care. But don't worry. You won't get bored to tears with some long essay on science and physics. We make the concepts and lingo easy to understand. Plus, in this part you'll find some great tips on safety. Electronics is fun, but only if you don't get burned, electrocuted, or poked in the eye by a wild resistor.

Chapter 1

From Electrons to Electronics

*W*hen you plug in the coffee maker in the morning, you're using electricity. When you flip on the TV to watch a rerun of *Sex in the City*, you're using electricity again (for better or worse).

You use electricity and electronics devices all the time, and you've finally worked up enough curiosity to want to tinker with electronic gadgets yourself. That's great. But before you can jump into playing with wires and batteries, it helps to understand what puts the *elec* in electricity and electronics.

In this chapter, you discover all about how electrons make electricity and how harnessing that electricity is the basis of electronics. You also get an introduction to some of the tools and parts that you can play with in the electronics projects in Chapters 14 and 15.

Just What Is Electricity?

Like most things in life, electricity is more complex than you may think. A lot of conditions have to come together to make that little spark when you touch a doorknob or provide the power to run a supercomputer. To understand how electricity works, it helps to break it down into its parts.

First, you take an electron

Electrons are one of the building blocks of nature. Electrons are buddies with another of nature's building blocks, protons. Electrons and protons are very small and are contained in . . . well, everything. A speck of dust contains millions and millions of electrons and protons, so you can imagine how many there are in your average sumo wrestler.

Electrons and *protons* have equal and opposite electric charges, with electrons having the negative charge and protons the positive. Opposite charges are attracted to each other. You can visualize a similar type of attraction by putting the ends of two magnets together. If the ends of the magnets are opposite poles, the magnets cozy right up to each other and stick together. If the ends of the magnets are the same pole, the magnets will move apart like two politicians in a heated debate. In a similar way, because electrons and protons have opposite charges, they are attracted to each other just as you can see opposite magnetic poles attracting. The attraction between electrons and protons acts like glue on a microscopic scale, holding matter together.

Although protons stay reasonably static, electrons are adventurous little fellows who don't like to just sit around at home. They can, and often do, move from one object to another. Walk across a carpet on a dry day and touch a doorknob; electrons traveling between your finger and the doorknob cause the spark that you feel and sometimes see. Lightning is another example of electrons traveling between two things — in this case, between a cloud and the ground. These examples both show electricity in an unharnessed state.

Moving electrons around through conductors

What do electrons use to travel from one place to another? The answer to that question gives you the next piece of the electricity puzzle. Although you may use your old Chevy to get around, electrons use something called a conductor. *Electricity* is simply the movement of electrons through a conductor.

A lot of materials can act as conductors, but some are much better at it than others. Electrons can move more easily through metal than through plastic. In plastic, even though all the electrons are moving around their proton buddies, they pretty much stay in their own backyard. But in metal, the electrons are free to move all over the place. Free electrons in metal act like marbles thrown on an ice-skating rink. The electrons glide through the metal like the marbles slide across the ice. Plastic, an insulator, is more like sand. Marbles don't go much of anywhere if you throw them into a sandbox, and neither do electrons in an insulator.

So which materials are good conductors and which are good insulators? Most folks use copper and aluminum as conductors. In fact, electronics projects often use copper wire conductors. Plastic and glass are commonly used insulators.

Resistance is the measurement of the ability of electrons to move through a material. A copper wire with a large diameter has lower resistance to the flow of electrons than a copper wire with a small diameter. You need to understand resistance because almost every electronics project you do involves a resistor. Resistors have controlled amounts of resistance, which allows you to control the flow of electrons in a circuit.

Voltage, the driving force

The previous sections in this chapter explain how electrons move and that they move more freely in a conductor. But some kind of force has to pull the electrons from one place to another. This attractive force between positive and negative charges is an electromotive force called *voltage*. Negative electrons move toward a positive voltage by way of a conductor.

Remember Ben Franklin's adventure flying a kite in a storm? The spark he produced that night gave him an understanding of how an electric current moves. In Ben's case, electrons traveled down the wet string, which acted as a conductor. (This was at least in part because the string was wet. Try this same stunt with dry string and it doesn't work nearly as well). The voltage difference between the negatively charged clouds and the ground pulled the electrons down the wet string.

Don't try Franklin's experiment yourself! By flying a kite in a storm, you're basically playing with lightning — which can effectively turn you into toast.

What happened to protons?

You may have noticed that we stopped talking about protons. Although you should understand the positive and negative charges in protons and electrons, we're focusing on electrons because they're more mobile than protons. In most cases, it is electrons, and their negative charges, that move through conductors and generate electricity. But in special cases, such as batteries, positive charges also move through conductors.

To explain this process, you also have to get into things called ions, atoms, electrochemical reactions, and maybe even the concept of holes as used in semiconductor physics. Because you don't need to understand these concepts to complete the projects shown in this book (or most hobbyist level projects), we'll leave the more complex physics to Einstein and keep our focus on electrons.

Conventional current versus real current

Early experimenters believed that electric current was the flow of positive charges. So they described electric current as the flow of a positive charge from positive to negative voltage. Much later, experimenters discovered electrons and determined that the flow of electrons in wires goes from negative to positive voltage.

The original convention is still with us today, — so the standard is to depict the direction of electric current in diagrams with an arrow that points opposite to the direction that electrons actually flow. *Conventional current* is the flow of a positive charge from positive to negative voltage and is just the reverse of real current.

An important combo: Electrons, conductors, and voltage

Say that you have a wire (a conductor), and you attach one of its ends to the positive terminal of a battery and the other end of the wire to the negative terminal of the battery. Electrons then flow through the wire from the negative to the positive terminal. This flow of electrons is referred to as an *electric current*. When you combine electrons, a conductor, and voltage you create an electric current in a form that you can use.

To help you picture how conductors and voltage affect the flow of electric current in a wire, think of how water pressure and pipe diameter affect the flow of water through a pipe. Here's how this analogy works:

✔ Increasing water pressure causes more water to flow through the pipe. This is analogous to increasing voltage, which causes more electrons to flow, producing greater electric current.

✔ Using a larger diameter pipe allows more water to flow through the pipe for a given amount of pressure. This is analogous to using wire with a larger diameter, which allows more electrons to flow for a given voltage, producing greater electric current.

Where Do You Get Electricity?

Electricity is created when voltage pulls an electric current through a conductor. But when you sit down and run a wire between a switch and a light, just where do you get the juice (the electricity) to power that light?

There are many different sources of electricity — everything from the old walking-across-a-carpet-and-touching-a-doorknob kind to solar power. But to make your life simple, this book takes a look at the three sources that you're likely to use for electronics projects: batteries, your wall outlet, and solar cells.

They just keep on going: Batteries

A battery uses a process called electrochemical reaction to produce a positive voltage at one terminal and a negative voltage at the other terminal. The battery creates these charges by placing two different metals in a certain type of chemical. Because this isn't a chemistry book, we don't get into the guts of a battery here — but trust us, this is essentially what goes on.

Batteries have two terminals (a *terminal* is just a fancy word for a piece of metal to which you can hook up wires). You often use batteries to supply electricity to devices that are portable, such as a flashlight. In a flashlight, the bulb has two wires running to the battery, one to each terminal. What happens next? Something like this:

- ✔ Voltage pulls electrons through the wire from the negative terminal of the battery to the positive terminal.
- ✔ The electrons moving through the wire pass through the wire filament in the light bulb, causing the bulb to light up.

Because the electrons move in only one direction, from the negative terminal through the wires to the positive terminal, the electric current generated by a battery is called *direct current,* or DC. This is in contrast to alternating current (AC) which is discussed in the following section, "Garden-Variety Electrical Outlets."

The wires on a battery must connect to both terminals. This setup allows electrons to flow from one terminal of the battery, through the bulb, and all the way to the other terminal. If the electrons can't complete this kind of loop between negative and positive, electrons don't flow.

Garden-variety electrical outlets

When you plug a light into an electrical outlet in your wall, you're using electricity that originated at a generating plant. That plant may be located at a dam or come from another power source, such as nuclear power. Or it may be fired by coal or natural gas. Because of the way electricity is generated at a power plant, the direction in which the electrons flow changes 120 times a second, making a complete turnaround 60 times a second. This change in electron flow is called *alternating current,* or AC.

When the change in electron flow makes a complete loop, it's called a *cycle.* The number of cycles per second in alternating current is measured in *Hertz,* abbreviated Hz. The example of a cycle in the previous paragraph is based on the fact that the United States uses a 60 Hertz standard frequency; some other countries use 50 Hertz as a standard, which means that the electrons change direction 100 times a second.

Electricity generated at a dam uses water to turn a coil of wire inside a huge magnet. One of the properties of magnets and wires is that when you move a wire near a magnet, a flow of electrons is induced in the wire. First, the magnet causes the electrons to flow in one direction, and then, when the wire loop rotates 180 degrees, the magnet pulls the electrons in the other direction. This rotation creates alternating current.

Just plugging a cord into a wall outlet sounds easy enough, but you need direct current for most projects, rather than alternating current. If you use wall outlets to supply electricity for your project, you have to convert the electricity from AC to DC. You can do this conversion with something called a power supply. For an example of a power supply, think of the charger that you use for your cell phone; this little device essentially converts AC power into DC power that the battery uses to charge itself back up. You can find out more about power supplies in Chapter 3.

Safety, safety, safety. It's an important issue for you to consider when deciding whether to use the AC electricity that you get from wall outlets. Using the electricity from a battery is like petting a house cat. Using the electricity from wall outlets is more like cozying up to a hungry lion. With a cute tabby, you may get your hand scratched; with the king of the jungle, you may be eaten alive. If you think that you need to use electricity from a wall outlet for a project, make sure that you know what you're doing first. See Chapter 2 for specific advice about safety.

Which came first, voltage or current?

Batteries produce a voltage that drives an electric current. Generators at dams drive a current that produces a voltage. Which comes first?

This is like asking yourself the well-known question about the chicken and egg. Voltage, currents, and conductors all work together. If there is a voltage applied across a conductor, electric current flows. If you have an electric current flowing through a conductor, there will be voltage across the conductor. Bottom line: Don't worry about which comes first.

A simple choice: AC or DC

What difference does it make to you if you use alternating or direct current? A lot of difference!

AC costs less to generate and send over transmission lines than DC. That's why you use AC for many household electricity needs, such as powering light bulbs and heaters.

However, DC is simpler to use for the projects discussed in this book (and many other electronics applications). It's just plain harder to control AC current because you don't know which way it's headed at any point in time. It's the difference between controlling traffic on a two-way, six-lane highway, and controlling traffic on a one-lane, one-way street. So, most of the circuits you read about in this book use direct current.

Solar cells

Solar cells are a form of semiconductor. Like batteries, solar cells have wires attached to two terminals. Shining light on a solar cell causes an electric current to flow. (This reaction to light is a property of semiconductors and is discussed in the sidebar "Getting fancy with semiconductors," later in this chapter.) The current is then conducted through wires to devices, such as a calculator or a garden light beside the pathway to your front door.

Using a calculator containing a solar cell, you can demonstrate that the calculator depends on the light shining on the solar cell for its power. Turn the calculator on and punch some numbers into the screen (choose a nice big number, like your income tax). Now, use your thumb to cover the solar cell. (The solar cell is probably near the top of the calculator in a rectangular area with a clear plastic cover.) After you've covered up the solar cell for a moment, the numbers fade away. Take your thumb off the solar cell, and the numbers reappear. Things powered by solar cells need light to work.

Where Do Electrical Components Fit In?

Electrical *components* are parts you use in electronics projects. Simple enough, right? You use some electrical components to control the flow of electricity, such as a dimmer switch that adjusts the brightness of a light. Electricity simply powers other electrical components, such as speakers blasting out sound. Still other electronic components, called *sensors,* detect something (such as light or heat) and then generate a current to do something in response, such as set off an alarm.

In this section, you meet some basic electrical components. Chapters 4 and 5 provide much more detail about components.

Controlling electricity

Electrical components, or parts, can control electricity. For example, a switch connects a light bulb to electric current. To disconnect the light bulb and make it go dark, the switch simply makes a break in the circuit.

Some other parts that control electricity are resistors, capacitors, diodes, and transistors. You can find more information on these parts in Chapter 4.

Controlling electricity even better (ICs)

Integrated circuits, or ICs, are components that contain a whole bunch of miniature components (such as resistors, transistors, or diodes, which you hear about in Chapter 4) in one device that may not be much bigger than an individual component. Because each IC contains many components, one little IC can do the same job as several individual parts.

Getting fancy with semiconductors

Transistors, diodes, LEDs, integrated circuits, and many other electronic devices use a semiconductor instead of a conductor. A *semiconductor* is a material, such as silicon, that has some of the properties of both conductors and insulators.

Silicon is pretty cool stuff. In fact, they've named a whole valley in California after it. In its pure state, silicon conducts an electric current poorly. But if you add contaminates, such as boron or phosphorus, to the silicon, it conducts. When you add phosphorus, silicon becomes an "n"-type semiconductor. When you add boron, silicon becomes a "p"-type semiconductor. An *"n"-type semiconductor* has more electrons than a pure semiconductor and a *"p"-type semiconductor* has fewer electrons than a pure semiconductor.

When the regions containing boron and phosphorus are next to each other in silicon, you have a *"pn" junction.* Current flows in only one direction across a "pn" junction. *Diodes,* components that can convert AC to DC by limiting the flow of current to one direction, are an example of a component that contains a "pn" junction.

A "pn" junction generates an electric current when exposed to light; this property is used when building solar cells. On the other hand, when you run an electric current through a "pn" junction, it emits light, as light-emitting diodes (LEDs) do.

Transistors use junctions in which three adjacent areas have contaminants added. For example, one region with phosphorus, one with boron, and another with phosphorus result in an "npn" junction. In a transistor, you apply a current to the middle of the three regions (the base), allowing a current to flow.

Most electronics projects you work on use components such as transistors, diodes, and integrated circuits, and these are made with semiconductors. It's semiconductors that have made possible much tinier electronic gadgets (like handheld computers and palm-sized radios).

An audio amplifier is one example of an IC. You can use audio amps to increase the power of an audio signal. For example, if you have a microphone, its small output signal is fed through an audio amplifier to make a strong enough signal to power a speaker.

Another type of IC used in electronics projects is a *microcontroller,* a type of integrated circuit that you can actually program to control cool gadgets like robots. We discuss microcontrollers in more detail in Chapter 13.

Sensing with sensors

Certain electrical components generate a current when you expose them to light or sound. You can use the current generated, together with a few of the components listed in the previous sections that control electricity, to turn on or off electronic devices, such as light bulbs or speakers.

Motion detectors, light sensors, microphones, and temperature sensors all generate an electrical signal in response to a stimulus (motion, light, sound, or temperature, respectively). These signals can then be used to turn other things on or off. A high signal level might turn something on and a low signal level turn something off. For example, when a salesperson walks up to your house, a motion detector can turn on a light (or better yet, sound a general alarm).

These signals take different forms, depending on the component supplying them. For example, a microphone supplies an AC signal, and a temperature sensor supplies a DC signal.

Figure 1-1 shows diagrams of a few signals that you run into often when working with electronics. These signals include

- ✔ **+ 5 Volt DC signal:** A high input.

- ✔ **0 volt DC signal:** A low input.

- ✔ **0 to 5 volt DC square wave:** The output of an *oscillator* (a device that cycles between high and low voltage); if you use this signal as input to a light bulb, it causes the light to blink on and off.

- ✔ **- 5 volt to + 5 volt AC sine wave:** A signal, such as from a microphone, that generates alternating current that a device, such as an amplifier, uses as input. A microphone generates the waveform in Figure 1-1 when it receives the sound produced by a tuning fork. Notice in Figure 1-1 that the transitions from +5 volts to -5 volts are gradual for the sine wave and more abrupt in the square wave.

You can find out more about various types of sensors in Chapter 5.

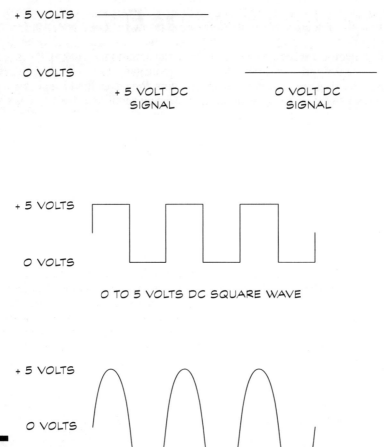

Figure 1-1:
Just a few
examples of
input
signals.

Powering up

Electricity can power electrical components to produce light, heat, sound, motion, and more. For example, an electric current supplied to a DC motor causes the shaft of the motor to rotate, along with anything you've attached to that shaft.

You can power speakers, light bulbs, LEDs, and motors with electricity. If you want to read more about these types of components, check out Chapters 4 and 5.

How Electricity Becomes Electronics

When you need to use electricity to make something work, such as a boom box, you've entered the world of electronic gadgets. No doubt you're eager to start making your own electronic gadgets. We cover the basics of how electronics and gadgets interact in the following sections.

Creating a simple circuit

Take a battery, a resistor, an LED, and some wires, put them together, and you have a simple electronic circuit. That's all an electronic *circuit* is — wires connecting components so that a current can flow through the components and back to the source.

Figure 1-2 shows a simple circuit. You place the parts in this circuit (also called components) on something called a breadboard and connect those parts with wires. If you've ever played with Mr. Potato Head, you understand the principle of a breadboard. You stick things in the potato (ears, a hat, eyes, and so on) to form a potato person. In the same way, a *breadboard* has slots for you to insert electronic components to build a sample circuit. If you're really happy with what you've created, you can then use that design to get a printed circuit board made. (See Chapter 11 for more information on building circuits on breadboards.)

Figure 1-2:
A collection
of parts is
assembled
into a
circuit.

Figure 1-2 shows wires connected to both terminals of the battery in the circuit. This connection allows the current to flow from the battery, through the LED and other components, and back to the battery to complete the circuit. You can also complete the circuit by connecting parts of the circuit to the metal chassis of a gadget, such as the metal housing of a stereo. We call this connection a *ground* because it is used as the reference for all voltages in the circuit. Ground may or may not be connected to the actual earth, but it is always the reference from which you measure all other voltages. We discuss grounding in detail in Chapter 6.

You can represent a circuit as a schematic. A *schematic* is just a drawing showing how components are connected together by wires. Check out the schematic for the circuit in Figure 1-2 in Figure 1-3. You can go to Chapter 6 for more on schematics.

Figure 1-3:
Can you
decipher
this
schematic
of the circuit
shown in
Figure 1-2?

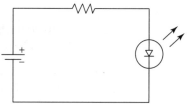

Deciding what to build

If you're itching to build a simple circuit to try out your skills, you can find several circuits in Chapter 14. For example, you can create a breadboard circuit that sounds an alarm when someone turns on a light in your room. Building these projects is a fun way to get familiar with how to put together a circuit. (But don't jump right into projects if you're a beginner — not until you've read through a few chapters in this book, especially Chapter 2 about safety.)

After you put together some of the breadboard projects in Chapter 11 and build up your basic skills, you can move on to the projects in Chapter 15, such as constructing a small robot. These projects take more time, but they can result in some truly neat gadgets.

After you've developed your skills building some of the projects in this book, you can go farther. One place to get additional ideas is on the Internet. Two sites we recommend are `discovercircuits.com/` and `www.electronics-lab.com`.

Along the Way You Get to Play with Tools

One of the best things about building electronics projects is that you get to tinker with tools and parts and see what you can make from them. You use some tools to put the circuits together and some tools to check out how the circuits you build are working.

Tools to build things

You're probably glad to hear that you don't need that many tools to get started. You just need a wire cutter, needle-noise pliers, a wire stripper, and a few screwdrivers to get started with the projects covered in Chapter 14.

If you design a circuit that you want to make more permanent, you need to get a soldering pencil (also called a soldering iron) to attach the elements of a circuit together. We cover choosing a soldering pencil in Chapter 8.

As you work with projects, no doubt other miscellaneous tools pop up that you may want to get your hands on. You can use a magnet to retrieve screws and other tiny things that you inevitably drop in hard-to-reach places, for example. Check out Chapter 3 for details on outfitting your workbench.

Tools to measure things

When building or troubleshooting a circuit, you need to make measurements to check that parts are working the way they should and that you designed and built the circuit correctly. Tools that you can use to measure things include a multimeter, an oscilloscope, and a logic probe. Chapters 9 and 10 cover the use of these tools.

We'll take a moment to briefly tell you what you can use a multimeter for because it's the measuring tool that you buy first and possibly the only one that you ever need.

Say you build a circuit, and you've just turned it on. What if the circuit doesn't work? With a multimeter, you can find out which part of the circuit is causing the problem. You can measure voltage, resistance, and current at different points on the circuit. For example, if there are 5 volts at one location on the circuit and further along at another location your voltage suddenly drops to 0 volts for no logical reason you can make a good guess that your problem lies between those two locations. You can then check (after the power is discon-nected, please!) for loose wires or damaged parts between those two locations.

Before troubleshooting a circuit for problems, read Chapter 2 on safety. You can very easily hurt yourself or your electronic gadget if you're not careful.

The Wonderful World of Units

To understand the results of your multimeter measurements, you need to understand electrical units. In the following sections, we run through the basics with you.

Measuring things in units

Units simply tell you how much of something you have. For example, when you buy apples, you measure how much they weigh in pounds (lbs). Similarly, a multimeter measures resistance in ohms, voltage in volts, and current in amperes (amps for short).

Table 1-1 shows common units and abbreviations used in electronics.

Table 1-1	Units Used in Electronics			
Term	*Abbreviation*	*Unit*	*Unit Symbol*	*Component*
Resistance	R	ohm	Ω	Resistor
Capacitance	C	farad	F	Capacitor
Inductance	L	Henry	H	Inductor
Voltage	E or V	volt	V	
Current	I	amp	A	
Power	P	watt	W	
Frequency	f	hertz	Hz	

Getting to bigger or smaller units

If you're measuring apples, you may have a tiny wedge of an apple (a fraction of an apple) or a few pounds of apples, right? Electronics has much larger ranges of units. You can have a single circuit using millions of ohms or another one with a very small current (maybe a thousandth of an amp). Talking about these very, very big numbers and very, very tiny numbers requires some special terminology.

Electronics uses things called prefixes and scientific notation to indicate small or large numbers. Table 1-2 shows common prefixes and scientific notations used in electronics.

Table 1-2	Prefixes used in Electronics			
Number	*Name*	*Scientific Notation*	*Prefix*	*Abbreviation*
1,000,000,000	1 billion	10^9	giga	G
1,000,000	1 million	10^6	mega	M
1,000	1 thousand	10^3	kilo	k
100	1 hundred	10^2		
10	ten	10^1		
1	one	10^0		
0.1	tenth	10^{-1}		
0.01	hundredth	10^{-2}		
0.001	1 thousandth	10^{-3}	milli	m
0.000001	1 millionth	10^{-6}	micro	μ
0.000000001	1 billionth	10^{-9}	nano	n
0.000000000001	1 trillionth	10^{-12}	pico	p

So how does this 10^{-6} or 10^6 stuff work? *Scientific notation* is basically a shorthand method of telling how many zeros to add to a number using our decimal system, which is based upon powers of 10. For example, the superscript '6' in 10^6 means place the decimal point six places to the right. 10^{-6} means move the decimal point six places to the left. So, with 1 x 10^6, you move the decimal point 6 places to the right of the 1, which gives you 1,000,000 or 1 million. With 1 x 10^{-6}, you move the decimal point 6 places to the left, giving you 0.000001 or 1 millionth. With 3.21 x 10^4, you move the decimal point 4 places to the right, for a result of 32,100.

Prefixes + units = ?

The previous section shows you the abbreviations for prefixes and units. This section tells you how to combine them. Combining these two results in very compact notation. For example, you can write 5 milliamps as 5 mA or 3 megahertz as 3 MHz.

Just as you usually use a pound or so of apples to bake your average pie or several tons of steel to build a suburban office park, in electronics, some things just naturally come in small measurements and others in large measurements. That means that you typically see certain combinations of prefixes and units over and over. Here are some common combinations of notations for prefixes and units:

- **Current:** pA, nA, mA, µA, A
- **Inductance:** nH, mH, µH, H
- **Capacitance:** pF, nF, mF, F
- **Voltage:** mV, V, kV
- **Resistance:** Ω, kΩ, MΩ
- **Frequency:** Hz, kHz, MHz, GHz

Exploring some new terms

Although we discussed resistance, voltage, and current earlier in this chapter, some other terms in this section may be new to you.

Capacitance is the ability to store a charge in an electric field. This stored charge has the effect of making decreases or increases of voltage more gradual. You can use components called *capacitors* to provide this property in many circuits. This figure shows the signal that occurs when you decrease voltage from +5 volts to 0 volts, both with and without a capacitor.

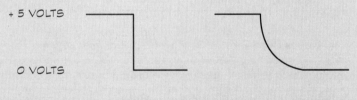

Frequency is a measurement of how often an AC signal repeats. For example, voltage from a wall outlet undergoes one complete cycle 60 times a second. The following figure shows a sine wave. In this figure, the signal completes one cycle when the current goes from -5 to +5 volts then back down to -5 volts. If a signal repeats this cycle 60 times a second, it has a frequency of 60 hertz.

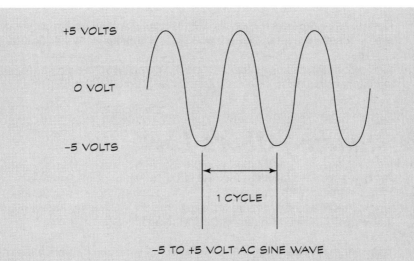

+5 VOLTS

0 VOLT

-5 VOLTS

1 CYCLE

-5 TO +5 VOLT AC SINE WAVE

Inductance is the ability to store energy in a magnetic field; this stored energy resists changes in current just as the stored charge in a capacitor resists changes in voltage. Components called *inductors* are used to provide this property in circuits.

Power is the measure of the amount of work that electric current does while running through an electrical component. For example, when voltage is applied to a light bulb and current is driven through the filament of the bulb, work is done in heating the filament. In this example, you can calculate power by multiplying the voltage applied to the light bulb by the amount of current running through the filament.

Using the information in Tables 1-1 and 1-2, you can translate these notations. Here are some examples:

- ✔ **mA:** milliamp or 1 thousandth of a amp

- ✔ **µV:** microvolt or 1 millionth of a volt

- ✔ **nF:** nanofarad or 1 billionth of a farad

- ✔ **kV:** kilovolts or 1 thousand volts

- ✔ **MΩ:** megohms or 1 million ohms

- ✔ **GHz:** gigahertz or 1 billion hertz

The abbreviations for prefixes representing numbers greater than 1, such as *M* for *mega,* use capital letters. Abbreviations for prefixes representing numbers less than 1, such as *m* for *milli,* use lowercase. The exception to this rule (there's always one) is *k* for *kilo,* which is lowercase even though it stands for 1,000.

The use of capital K is a special case reserved for kilohms; when you see a capital K next to a number such as 3.3k, this translates as 3.3 kilohms.

You have to translate any measurement expressed with a prefix to base units to do any calculation, as you can see in the following sections.

Understanding Ohm's Law

Say that you're wiring a circuit. You know the amount of current that the component can withstand without blowing up and how much voltage the power source applies. So you have to come up with an amount of resistance that keeps the current below the blowing-up level.

In the early 1800s, George Ohm published an equation called Ohm's Law that allows you to make this calculation. *Ohm's Law* states that the voltage equals current multiplied by resistance, or in standard mathematical notation

$$V = I \times R$$

Taking Ohm's Law farther

Remember your high school algebra? Remember how if you know two things (such as *x* and *y*) in an equation of three variables, you can calculate that third thing? Ohm's Law works that way; you can rearrange its elements so that if you know any two of the three values in the equation, you can calculate the third. So, here's how you calculate current: current equals voltage divided by resistance, or

$$I = \frac{V}{R}$$

You can also rearrange Ohm's Law so that you can calculate resistance if you know voltage and current. So, resistance equals voltage divided by current, or

$$R = \frac{V}{I}$$

So far, so good. Now, take a specific example using a circuit with a 12-volt battery and a light bulb (basically, a big flashlight). Before installing the battery, you measure the resistance of the circuit with a multimeter and find that it's 9 ohms. Here's the formula to calculate the current:

$$I = \frac{V}{R} = \frac{12 \text{ volts}}{9 \text{ ohms}} = 1.3 \text{ amps}$$

What if you find that your light is too bright? A lower current reduces the brightness of the light, so just add a resistor to lower the current. Originally, we had 9 ohms; adding a 5-ohm resistor to the circuit makes the total resistance 14 ohms. In this case, the formula for current is

$$I = \frac{V}{R} = \frac{12 \text{ volts}}{14 \text{ ohms}} = 0.9 \text{ amps}$$

Dealing with numbers both big and small

Say that you have a circuit with a buzzer that has resistance of 2 kilohms and a 12-volt battery. You don't use 2 kilohms in the calculation. To calculate the current, you have to state the resistance in the basic units, without using the "kilo" prefix; in this example that means that you have to use 2,000 ohms for the calculation, like this:

$$I = \frac{V}{R} = \frac{12 \text{ volts}}{2,000 \text{ ohms}} = 0.006 \text{ amps}$$

You now have the calculated current stated as a fraction of amps. After you finish the calculation, you can use a prefix to restate the current more succinctly as 6 milliamps or 6 mA.

Bottom line: You have to translate any measurement expressed with a prefix to base units to do a calculation.

The power of Ohm's Law

Ohm (never one to sit around twiddling his thumbs) also expressed that power is related to voltage and current using this equation:

P = V x I; or power = voltage x current

You can use this equation to calculate the power consumed by the buzzer in the previous section:

P = 12 volts x 0.006 amps = 0.072 watts which is 72 milliwatts (or 72 mW)

What if you don't know the voltage? You can use another trick from algebra. (And you thought Mrs. Whatsit wasted your time in Algebra 101 all those years ago!) Because V = I x R, you can substitute I x R into this equation, giving you

$P = I^2 \times R$; or power = current squared x resistance

You can also use algebra to rearrange the equation for power to show how you can calculate resistance, voltage, and current if you know power and any one of these parameters.

Do you really hate algebra? Did Mrs. Whatsit fail you those many years ago? You're probably happy to hear that online calculators can make these calculations much easier. Try searching on www.google.com using the keyword phrase "Ohm's Law Calculator" to find them. Also, check out Chapter 18. It provides ten of the most commonly used electronics calculations.

Chapter 2

Keeping Humans and Gadgets Safe

You probably know that Benjamin Franklin "discovered" electricity in 1752 by flying a kite during a lightning storm. But actually, Franklin already knew about electricity. He was really just testing a form of lightning conductor. Though his experiment proved modestly successful, it was anything but safe. Franklin almost killed himself, and if he had, whose picture would be on the $100 bill?

Respect for the power of electricity is necessary when working with electronics. In this chapter, we take a look at keeping yourself — and your electronic projects — safe. This is the one chapter that you really should read from start to finish, even if you already have some experience in electronics.

The Sixth Sense of Electronics

The sixth sense of electronics isn't about seeing dead people. In this case, the sixth sense is common sense, the smarts that help you stay among the living. Common sense is that voice inside you that tells you not to stick your fingers in an empty lamp socket without first unplugging the lamp.

No book can ever teach you common sense. You have to cultivate it like an exotic flower. But a few words to the wise may help get you started in your quest for electronics common sense. For starters:

- ✔ **Never assume. Always double-check.** Pretend that your soldering pencil is out to get you. Your family may think you're crazy, but you're less likely to burn or electrocute yourself.

- ✔ **If you're not sure about how to do something, read up on it first.** Not everything in electronics is as obvious as it first appears.

- ✔ **Don't take chances.** If you can make a 50/50 bet that something is plugged in, give some thought as to what happens if you lose the bet this time.

Never let your guard down. Don't ruin the fun of a wonderful hobby or vocation because you neglected a few basic safety measures.

The Dangers of Electrical Shock

By far, the single most dangerous aspect of working with electronics is the possibility of electrocution. *Electrical shock* results when the body reacts to an electrical current — this reaction can include an intense contraction of muscles (namely, the heart) and extremely high heat at the point of contact between your skin and the electrical current. The heat leads to burns that can cause death or disfigurement. Even small currents can disrupt your heartbeat.

The degree to which electrical shock can harm you depends on a lot of factors, including your age, your general health, the voltage, and the current. If you're over 50 or in poor health, you probably won't stand up to injury as well as if you're 14 and as healthy as an Olympic athlete. But no matter how young and healthy you may be, voltage and current can pack a wallop, so it's important that you understand how much they can harm you.

Electricity = voltage + current

To fully understand the dangers of electrical shock, you need to know the basics of what makes up electricity. In Chapter 1, we state that electricity is made up of two elements: voltage and current.

Voltage and current work hand-in-hand and in ways that directly influence the severity of electrical shock. Consider the analogy of water flowing through a pipe. Think of the water as representing the electricity. Increasing the diameter of the pipe to let more water through is like increasing current. Imagine being under a drain pipe for the Hoover Dam! Increasing the pressure of the water in the pipe is like increasing voltage. You know that even small amounts of water at high pressure can be devastating. The same is true of electricity, where even low voltages at high currents can potentially kill you.

Is it AC or DC?

You can describe electrical current as being either of the following

- **Direct (DC):** The electrons flow one way through a wire or circuit.
- **Alternating (AC):** The electrons flow one way, then another, in a continuing cycle.

If this stuff is new to you, you may want to go back and read Chapter 1 for a more detailed discussion.

Household electrical systems in the U.S. and Canada operate at about 117 volts AC. This significantly high voltage can, and does, kill. You must exercise *extreme caution* whenever you work with it.

Until you become experienced working with electronics, you're best off avoiding circuits that run directly off household current. Stay with circuits that run off standard-size batteries, or those small plug-in wall transformers. Unless you do something silly, like lick the terminal of a 9-volt battery (yes, you get a shock!), you're fairly safe with these voltages and currents.

The main danger of household current is the effect it can have on the heart muscle. High AC current can cause severe muscle contraction, serious burns, or both. And many electrocution accidents occur when no one is around to help the victim.

Burns are the most common form of injury caused by high DC current. Remember that voltage doesn't have to come from a souped-up power plant to be dangerous. For example, don't be lulled into thinking that because a transistor battery delivers only nine volts, it's harmless. If you short the terminals of the battery with a piece of wire or a metal coin, the battery may overheat — and can even explode! In the explosion, tiny battery pieces can fly out at high velocity, burning skin or injuring eyes.

Trying to not get electrocuted

Most electrocution accidents happen because of carelessness. Be smart about what you're doing, and you will significantly reduce the risk of being hurt by electricity.

Here are a few handy electrocution prevention tips:

✔ **Avoid working with AC-operated circuits.** Of course, you can't always do this. If your project requires an AC power supply (the power supply converts the AC to lower-voltage DC), consider using a self-contained one, such as a plug-in wall transformer. They're much safer than a home-made power supply.

✔ **Physically separate the AC and DC portions of your circuits.** This helps to prevent a bad shock if a wire comes loose.

✔ **Make sure you secure all wiring inside your project.** Don't just tape the AC cord inside the project enclosure. It may pull out sometime, exposing a live wire. Use a strain relief or a cable mount to secure the cord to the enclosure. A *strain relief* clamps around the wire and prevents you from tugging the wire out of the enclosure. You can buy a strain relief for electrical cords at almost any hardware store or electronics shop.

✔ **Whenever possible, use a metal enclosure for your AC-operated projects, but only if the enclosure is fully grounded.** You need to use a 3-prong electrical plug and wire for this. Be sure to firmly attach the green wire (which is always the ground wire; ground is used as a reference for all voltages in a circuit) to the metal of the enclosure.

✔ **If you can't guarantee a fully-grounded system, use a plastic enclosure.** The plastic helps insulate you from any loose wires or accidental electrocution. For projects that aren't fully grounded, only use an isolated power supply, such as a wall transformer (a black box with plug prongs which is attached to a wire, such as you may have on your cell phone charger). You plug the transformer into the wall, and only relatively safe low-voltage DC comes out.

✔ **Don't be the class clown.** Be serious and focused while you're working around electricity.

✔ **Don't work where it's wet.** "Yeah, duh!" you say. But you'd be surprised what people sometimes do when they're not paying attention. And remember, just because you put liquid in a cup, that doesn't mean you don't run the risk of knocking it over and getting things wet; consider leaving your soft drink or coffee on an out of the way shelf when working on your electronics project.

Practice the buddy system. Whenever possible, have a buddy nearby if you're working around AC voltages. You want someone around who can dial 9-1-1 when you're lying on the ground unconscious. Seriously.

Getting a first aid chart

Of course, you're the safest person on earth, and you will never be electrocuted. But just in case, get one of those emergency first aid charts that includes information about what to do if anyone else (not you, of course) ever pokes his finger into a wall outlet. You can find these charts on the Internet; try a search for "first aid wall chart." You can also find them in school and industrial supply catalogs.

Helping someone who has been electrocuted may require cardio-pulmonary resuscitation, otherwise known as CPR. Be sure that you're properly trained before you administer CPR on anyone. Otherwise, you may cause more harm than good. Check out *www.redcross.org* to get more information about CPR training.

Zaps, Shocks, and Static Discharge

One type of everyday electricity that is dangerous to both people and electronic gizmos is static electricity. They call it *static* because it's a form of current that remains trapped in some insulating body, even after you remove the power source. With conventional AC and DC current, static electricity disappears when you turn off the power source.

The ancient Egyptians discovered static electricity when they rubbed cat fur against the smooth surface of amber. After they rubbed the materials together, they tended to cling to one another by some unseen force. Similarly, two pieces of cat fur that they rubbed against the amber tended to separate from each other when the Egyptians drew those pieces together. Although the Egyptians didn't understand this mysterious force, they were aware of it. And they had the scratched-up arms to prove it! (Note to Pharaohs: Best not to use live cats for your electricity experiments.)

Carpets don't shock, people do

Carpet shock hasn't killed anyone (that we know of, anyway). The amount of current is usually too low to harm your body. But, because of their extremely small size, the same isn't true for electronic components. Static electricity of just a few thousands volts, a mere tingle to you (because the current is so very, very low), can cause great harm to sensitive electronic components.

As an electronics experimenter, remember to take specific precautions against electrostatic discharge, or ESD. See the section "Tips for reducing static electricity," later in this chapter, for specific pointers. You can all but eliminate damage from static discharge by taking just a few simple steps to protect yourself, your tools, and your projects from static buildup. The cost for protecting against static electricity is minimal; without knowing it, you may already be on the road to preventing dangerous static buildup in your electronics workshop.

Static electricity hangs around until it dissipates in some way. Most static dissipates slowly over time, but in some cases, it gets released all at once. Lightning is one of the most common forms of static electricity.

Designers make certain common electronic components to hold a static charge, such as the ordinary capacitor (a component that provides the ability to store energy in an electric field). Most capacitors in electronic circuits store a very minute amount of charge for extremely short periods of time. But some capacitors, most notably those used in bulky power supplies, can store near-lethal doses for several minutes or even hours. *Use caution when working around capacitors that can store a lot of charge so you don't get an unwanted shock.*

That guy from the $100 bill again

Benjamin Franklin, like other scholars of his time, understood quite a bit about static electricity. One of Franklin's many inventions was an early motor that ran completely on static electricity. While Ben's motor is little more than a scientific curiosity today, it shows that static is a form of electricity, just like AC or DC electricity.

Imagine a motor without a battery. Ben Franklin had to imagine this because batteries weren't invented until after he died. The honor of inventing the battery in the year 1800 goes to Alessandro Volta — hence the name of the unit of measure for electromotive force (an attractive force between positive and negative charges), the volt . And even though Franklin didn't come up with the battery, he first coined the term to describe his apparatus that collected static in charged glass plates.

You can encounter static electricity now and then by doing nothing more than walking across a carpeted floor. As you walk, your feet rub against the carpet and your body takes on a static charge. Touch a metal object, such as a doorknob or a metal sink, and the static quickly discharges from your body. You feel that discharge as a slight shock.

How static can turn components to lumps of coal

Electrostatic discharge involves very high voltages at extremely low currents. Combing your hair on a dry day can develop tens of thousands of volts of static electricity, but the current is almost so negligible you seldom notice it. The low current prevents the static discharge from really hurting you when you receive a shock. Instead, you just get an annoying tickle (and maybe a bad hair day).

Many components that you use in electronic equipment, from simple transistors to complex integrated circuits, are quite sensitive to even low amounts of electrostatic discharge. Transistors and integrated circuits can be particularly sensitive to high voltages, regardless of the amount of current. These components include CMOS transistors, integrated circuits, and most computer microprocessors. Other electronic components also are sensitive to very high levels of electrostatic discharge, but you don't normally encounter these levels in everyday life. (You can read more about CMOS, transistors, and other components in Chapter 4.)

Not all electrical components are static sensitive, but for safety's sake, develop static-safe work habits for all the components that you handle. Table 2-1 lists the major electronic components and how susceptible they are to damage from static discharge. Read Chapters 4 and 5 for more information on what these components do.

Table 2-1	Static Sensitivity of Common Components	
Low	*Medium*	*High*
resistors	bipolar transistors	CMOS transistors and integrated circuits
capacitors	TTL integrated circuits	MOSFETs
diodes	many linear integrated circuits	microprocessors and related components
transformers		
coils		
all passive components, such as batteries, switches, and connectors		

Tips for reducing static electricity

You can bet that most of the electronic projects you want to build contain at least some components that are susceptible to damage from electrostatic discharge. You can take a number of simple steps to prevent exposing your projects to the dangers of electrostatic discharge:

  ✔ **Use an anti-static mat.** An anti-static mat acts to reduce or eliminate the chance of static building up on your table and yourself as you work with an electronic device. You can find anti-static mats in both table-top and floor varieties. The table-top mats look like a sponge, but it's really conductive foam. You can (and should) test the conductivity of the mat by placing the leads of a multimeter (a piece of testing equipment you can read more about in Chapter 9) on either side of a length of the mat. Dial the meter to ohms. You should get a definite reading and not an infinitely open circuit (a circuit that has a break in it; see Chapter 7 for more about circuits).

✔ **Use an anti-static wrist strap.** As a further aid in reducing static electricity, also wear an anti-static wrist strap when working on electronic gear. This wrist strap, like the one shown in Figure 2-1, grounds you at all times and prevents static build-up. This strap is one of the most effective means of eliminating electrostatic discharge — and it's one of the least expensive. Most anti-static wrist straps cost under $5 and are worth every penny. To use the strap, roll up your shirt sleeves and remove watches, bracelets, rings, and any other metallic objects. Wrap the strap around your wrist and make sure that it's tight. Securely attach the wire from the wrist-strap to a proper earth ground, as the instruction sheet that comes with the strap explains.

✔ **Wear low-static clothing.** Your choice of clothing can affect the amount of static build-up in your body. Whenever possible, wear natural fabrics, such as cotton or wool. Avoid wearing polyester and acetate clothing because these fabrics have a tendency to develop a whole lot of static. A cotton lab overcoat not only looks trendy (in that geeky sort of way), but it can reduce static electricity. Many chemical and industrial supply houses sell lab coats for reasonable prices. You also can find suitable overcoats, smocks, and aprons at many hardware stores.

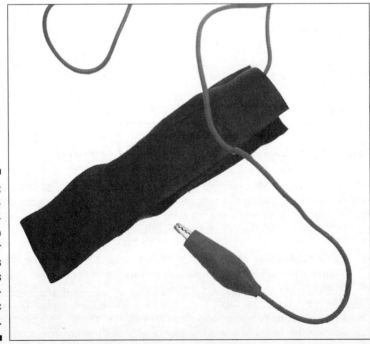

Figure 2-1:
An anti-static wrist-strap reduces or eliminates the dangers of electro-static discharge.

Grounding your tools

The tools you use when building electronics projects can also build up static electricity. A lot of it, in fact. If your soldering pencil operates from AC current, ground it as a best defense against electrostatic discharge. There's a double benefit here: A grounded soldering pencil not only helps prevent damage from electrostatic discharge but also lessens the chance of a bad shock if you accidentally touch a live wire with the pencil.

Cheapo soldering pencils use only two-prong plugs and don't have a ground connection. You can't find a really safe and sure means of attaching a grounding wire to the soldering pencil, so the best bet is to just buy a new and better pencil. You can purchase a grounded soldering pencil for less than $30, including an assortment of tips.

As long as you ground yourself by using an anti-static wrist strap, you generally don't need to ground your other metal tools, such as a wire wrapping tool, screwdrivers, and wire cutters. Any static generated by using these tools is dissipated through your body and into the anti-static wrist strap.

Working with AC Current

The vast majority of hobby electronics projects run on batteries. Simple enough, but some projects need more current or higher voltages than batteries can easily provide. Instead of building a power supply that converts household AC current to a DC voltage for your project, you can make things much safer for yourself by using a wall transformer to convert AC to DC (see Figure 2-2). All the working parts are self-contained in the wall transformer. As long as you don't try to take it apart, you don't expose the AC house current.

Where to get wall transformers — cheap!

You can purchase wall transformers (called "wall warts" by some because they stick out of the wall like an ugly wart) new or as surplus. Try Radio Shack or a similar electronics store to buy new wall transformers. You can get used and surplus wall transformers by mail-order surplus; check out Chapter 17 for some good leads.

And, of course, you may have a wall transformer saved from a discarded cordless phone or other electronic gadget. Check the voltage and current rating, usually printed on the transformer, to see if it's suitable for your next project.

Figure 2-2:
A plug-in
wall
transformer
shields you
from
exposed AC
household
current.

Sometimes, you need to work on a project that uses your 117 volt AC house current directly. In those cases, you can't resort to relatively safer batteries. No hiding behind a wall transformer either. For these projects, always exercise caution. Although you're being super careful, you can further minimize the hazards of working with circuits powered by AC house current by following these basic guidelines:

✔ **Always keep AC circuits covered.** A little sheet of plastic works wonders.

✔ **Never circumvent any fuse protection used on the device.** Don't use a fuse with a too-high rating and don't bypass the fuse altogether.

✔ **When troubleshooting AC circuitry, keep one hand in your pocket at all times.** This prevents you from accidentally touching things with your hand that you shouldn't. Use the other hand to manipulate the testing apparatus. Avoid the situation where one hand touches ground and the other a live circuit. The AC can flow from one hand to the other, straight through your heart.

✔ **If possible, use the buddy system when working with AC circuits.** Always have someone nearby who can help you in case you get a nasty shock.

✔ **Double- and triple-check your work before applying power.** If possible, have someone who knows a little about circuits inspect your handiwork before you switch the circuit on for the first time.

✔ **Periodically inspect AC circuits for worn, broken, or loose wires and components and make any necessary repairs.**

When testing AC-operated circuits, first remove the power. Unplug the power cord, don't just switch off the power at an outlet strip. You can easily tell when you've pulled the plug from the socket, but it's harder to tell if those little electrons are still swirling around the outlet strip.

The Heat Is On: Safe Soldering

When soldering, you use a hot soldering pencil or gun, working with temperatures in excess of 700 degrees Fahrenheit. To get an idea of what that temperature means, it's the same as an electric stove burner set at high heat. You can imagine how much that hurts if you touch it.

Most electronic projects or fix-it jobs call for a soldering pencil rather than those big soldering guns that look like they're rejects from a Buck Rogers movie. Chapter 8 discusses soldering in more detail, but for now, keep the following safety tips in mind:

✔ **Always place your soldering pencil in a stand designed for the job.** Never place the hot soldering pencil directly on a table or workbench. You can easily start a fire or burn your hands that way.

✔ **Be sure that the electrical cord doesn't snag on the table or any other object.** Otherwise, the hot soldering pencil can get yanked out of its stand and fall to the ground. Or worse, right into your lap!

✔ **Soldering produces mildly caustic and toxic fumes.** Make sure that your electronics workshop has good ventilation to prevent a buildup of these fumes. Avoid hunching over the soldering work because the fumes can waft into your face. Yuck. If you're having trouble seeing the soldering joint at a distance, use a magnifying glass to enlarge the image of the work.

✔ **If your soldering pencil has an adjustable temperature control, dial the recommended setting for the kind of solder that you're using.**

✔ **If you're concerned about stunting your growth and other health issues, you may want to avoid solders that have lead in them.** As an alternative, you can use lead-free rosin-core solder specifically designed for use on electronic equipment. Never use silver solder or acid-flux solder in electronics, by the way. They wreck your circuits.

✔ **Don't try to solder on a *live circuit* — a circuit to which you've applied voltage.** You run the risk of damaging the circuit or the soldering pencil, and you may receive a nasty shock.

✔ **Never grab a soldering pencil as it falls to the ground.** Just let it hit, and buy a new one if the pencil is damaged. There's an unwritten Murphy's Law in electronics that you will always grab the hot end. Trust me, a burn from a hot soldering pencil is something you don't ever want to experience.

Wearing Body Armor

Okay, so we may sound like overprotective mothers advising you to bundle up against the winter chill, but in the interest of practicing safe electronics, be sure to wear proper clothing and body part protectors while you work. Here are some specifics:

✔ Wear ear protection when using a high-speed drill or similar tool. Over time, the loud noise of the motor can harm your hearing.

✔ Wear protective glasses when assembling circuits, especially during soldering (this measure keeps out the solder fumes that can irritate your eyes) and when cutting wires. You don't want little wire bits flying into your eyeballs, after all.

✔ Wear comfortable clothing, but avoid anything dangling or loose-fitting. Roll up your sleeves, tuck in your shirttail, and if you're the formal type, remove your tie.

Avoid wearing metal jewelry when working around dangerous voltages. The metal can burn you if you're ever a victim of electrical shock. You probably don't need to worry about a ring, but you may want to reconsider that solid gold necklace.

Part II

Aisle 5,
Component Shack:
Stocking Up

The 5th Wave   By Rich Tennant

In this part . . .

This is where you'll learn what bits and pieces you need to collect before you can build your first electronic circuit. The chapters start with setting up shop: the tools you need, some suggestions for storing all your junk, and setting up your electronics workbench. Then you'll learn about several dozen electrical components — stuff like resistors, capacitors, and transistors — that you use in most typical electronics projects. You discover what they do, how they're used, and how to tell which is what. Or maybe that's what is which.

Chapter 3

Outfitting Your Electronics Bench

In This Chapter

▶ Exploring the basic hand tools that you use almost every day

▶ Getting your hands on some of the fun, non-essential tools

▶ Using cleaners, lubricants, and other critical chemicals

▶ Sticking things together with tape, glue, and other adhesives

▶ Finding a workspace and making it work for you

*F*orget all this voltage, current, and resistor stuff. You probably want to get down to the *really* fun part of electronics — the tools!

Every hobby has its special assortment of tools and supplies. Electronics is no exception. From the lowly screwdriver to the high-speed drill, you enjoy playing with electronics much more if you have the right tools.

You may have some or all of these tools already. If you do, you're ahead of the game. Gather them up, toss them into a toolbox, and skip on to the next chapter. But if you're tool-challenged in any way, don't let it get you down. You don't have to own every tool discussed in this chapter, and you can collect the ones you want as you go.

By the way, this chapter isn't totally comprehensive. It doesn't discuss soldering tools or test equipment, for example. You can read more about tools for soldering in Chapter 8. We cover test equipment, such as multimeters, logic probes, and oscilloscopes, in Chapters 9 and 10. And finally, you can find out about some tools specifically geared toward constructing printed circuit boards in Chapter 11.

Oh, the Hand Tools You Will Use

Hand tools are the mainstay of any toolbox. These tools tighten screws, snip off wires, bend little pieces of metal, and do all those other mundane tasks.

The following sections outline the basic hand tools that you need and what you need them for.

Screwdrivers (the tool, not the cocktail)

Unless you were brought up in the woods by wolves, you know what a screwdriver is. You use screwdrivers to put things together and take them apart with screws. Screwdrivers come in all sorts of sizes. You use fairly small ones for electronics. You may find a set of tiny, so-called "jeweler's screwdrivers" particularly handy for all the miniature stuff that you work with in electronics.

You can get screwdrivers just about anywhere, including hardware stores and discount stores, such as Wal-Mart. Buy them in sets to save money.

Driving Miss Phillips

Screws come with different types of drive heads, such as Phillips heads with their little (plus) + shape and slotted with a single score in the head (see Figure 3-1). You need to use a screwdriver that matches your screw head.

Be sure to use the correct size of screwdriver for any drive style. This tip is especially important when you use Phillips and specialty screws. Each drive style comes in several different sizes, and using the wrong size screwdriver can damage the head of the screw. So you may find it handy to buy an assortment of screwdrivers — that way, you're sure to have the right one when you need it.

Different screws for different jobs

Why the heck do screws have different types of drive heads? No one really knows for sure, but it may have to do with crop circles made by alien visitors to Earth. No, just kidding! Each drive type has its own advantages, depending on the application. Here's a quick rundown:

✔ Most screwdriver-using folks prefer slotted screws for general use because they work with a wider variety of blade sizes. (Even so, only one blade size is absolutely correct for any given screw head.)

✔ Phillips screws are easier to use in automated and semi-automated production. The screwdriver naturally slips into the screw's

slots and makes positive contact with the screw head. This certainty makes Phillips-head screws perfect for manufacturing lines, and most of the electronics gadgets, toys, and other products that you buy use them.

✔ Hex and specialty screw heads provide a positive, no-slip drive between screwdriver and screw. You want to use these heads when the screw has to be really tight, such as with the assembly of a high-speed machine or one that gets jostled a lot, like your car.

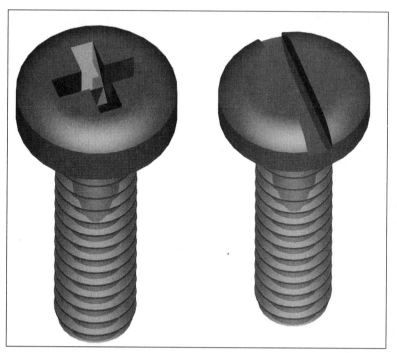

Figure 3-1:
Two screw
head styles.

How many types of screws can there be, anyway? Here's a list:

- **Slotted screws:** Probably the most common of all, these screws have a single slot. Use with a flat-blade screwdriver.

- **Phillips screws:** These screws have a plus (+) shaped slot. Use with a Phillips screwdriver. After slotted screws, you probably come across Phillips screws most often.

- **Hex screws:** These screws have a hexagon-shaped socket. Use with a hex screwdriver or a set of L-shaped hex wrenches. (You may hear these tools also called Allen or key wrenches.) No matter what specific tool you use, with a hex screw it *must* be the right size!

- **Specialty screws:** These screws use a variety of slot styles. Manufacturers make many of these screws for specific projects or distributors, and you don't see them often. They go by names like Torx and Pozi-Drive. Don't bother buying specialty screwdrivers until you need them. Like hex screws, you need to exactly match the specialty screw with the right screwdriver.

Screwdrivers with a magnetic personality

When working with small screws, having a magnetized screwdriver really helps. You can then use the screwdriver to pick up the screw and align the screwdriver (with the screw magnetically stuck to the end) with the hole or slot — all with one hand and no cussing! If you don't already have magnetized screwdrivers, purchase a screwdriver magnetizer at the hardware store. The magnetizer lets you magnetize and demagnetize your screwdrivers and other metal tools.

Not all screws are metal. Some are made of nylon or another plastic material, so obviously magnetizing your screwdrivers doesn't help much with these screws. Even some metal screws are non-magnetic, so your magnetic screwdriver doesn't have any supernatural powers over those little guys. These non-magnetic metal screws are often made of brass, aluminum, or some type of stainless steel.

Here's a trick if you're using non-magnetic screws and can't seem to hold them in position. Get a small package of rubber holdup putty, available at any office supply store. Pull off a very small portion of the putty and cram it into the head of the screw. Insert the screwdriver into the screw head. The screw should stay attached to the screwdriver long enough for you to start screwing it into the hole.

Take it off: Wire cutters and strippers

The wire cutter and stripper tool is a must-have for any electronics work. As the name suggests, you use the wire cutter and stripper to both cut wire and strip off the wire's plastic insulation. You can see a combination cutter and stripper in Figure 3-2. Look for these tools at Radio Shack and other electronics parts stores, or check out one of the better-stocked hardware and home improvement outlets.

With many strippers, you can "dial in" the gauge of the wire. (See the sidebar titled "What the Heck Is Wire Gauge, Anyway?" if you want to know about wire gauge.) This tool allows you to more easily remove just the insulation, without cutting or nicking the wire underneath.

What the heck is wire gauge, anyway?

You measure wire thickness in *gauge* or *AWG* (which stands for American Wire Gauge, if you're interested). The smaller the gauge, the larger the wire. The smallest wire commonly used in electronics is 30 gauge, for constructing wire wrap circuit boards (see Chapter 12 for more info on this technique). You should use wire strippers especially made for this small size.

Hookup wire for general electronics work is 20 to 22 gauge. You use this size for most projects. For heavier-duty applications, like wiring up large motors, you may use 16- or 18-gauge wire. As a point of reference, 20-gauge wire measures 0.032 inches in diameter, whether it's made of one solid conductor or a twine of many conductors twisted together. You read more about wire and wire gauge in Chapter 5.

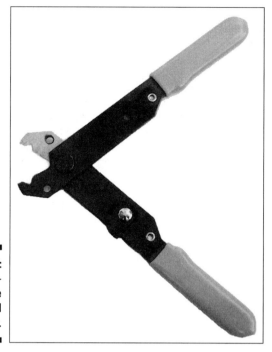

Figure 3-2:
A combination wire cutter and stripper.

Some folks prefer to buy wire cutters and wire strippers as separate tools. Either the cutter or stripper tends to get dull faster than the other (depending on the type of work that you do and the wire that you use). The separate tools tend to be cheaper than one of the combo deals, and you don't have to pay as much down the road if you replace them one at a time.

Another form of wire cutter is the flush or nippy cutter, which you can see in Figure 3-3. It cuts flush with a printed circuit board, and is useful when you need to get in close. The tool works well with wire from 30 to 16 gauge. Thicker wire may damage the tool or dull the cutting blades. For thicker wires, use diagonal cutters, also called lineman's pliers.

Getting a grip with needle-nosed pliers

Pliers help you grip stuff, bend wires, and hold parts in place during project assembly. For intricate work, use a 5-inch, needle-nosed pair of pliers, such as those that jewelry makers use. You can use larger pliers for general-purpose work. (By the way, the size of the pliers reflects the overall dimensions of the tool, not how large the jaws open.)

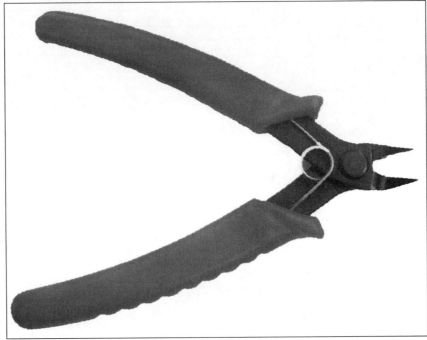

Figure 3-3:
Nippy
cutters trim
wire ends
flush to the
surface.

Be sure that you use the proper size pliers for the job. Using pliers that are too small may ruin the tool. And using a tool that's too large may damage components or cause unnecessary frustration.

Magnifiers: The better to see you with

A 4X to 8X magnifying glass helps you zoom in close and inspect your handi-work. You may find the magnifying glass particularly handy when you're look-ing for solder bridges, cold solder joints, or incomplete joints. (We cover these soldering gotchas more fully in Chapter 8.)

4X or *8X* means the magnifier enlarges the image by four or eight times, respectively. You can get magnifiers with other magnification powers. Anything less than 4X may not enlarge the image enough to be of any use to you, and anything more than 8X may be too powerful to do the kind of detail work that electronics requires.

Take a look at the magnifying glass in Figure 3-4. It's attached to a set of adjustable clips that hold small parts while you're working. You can find this type of rig, called "helping hands" or "third hand," particularly useful when you're soldering or at any other time that you need to work with small parts.

Another option for you to consider is magnifying glasses that you wear on your head. Sounds painful, but it's not. The unit slips over your head like a baseball cap, with the magnifying glasses positioned in front of your eyes. You can flip the magnifiers out of the way when you don't need them.

Figure 3-4:
These helping hands combine alligator clips with a magnifying glass.

A place for everything, and everything in its place

Over the months and years that you play with electronics, you can amass quite a few bits and pieces of junk. You want to keep track of it all, and you can do this bit of organization easily using a parts bin, also affectionately called a junk box. These bins have drawers for storing nuts, screws, resistors, capacitors, and other little parts. Choose the bin that has the number and size of drawers that you want. I like the type with both small and large drawers; the large drawers accommodate the bigger parts, as well as some tools and supplies, such as solder.

You may find that making labels to mark what's in each drawer is really useful. You can hand-write labels or use a labeler machine, such as the Brother P-Touch. For drawers that hold several different things, you can use dividers that keep parts separated. Provide a separate label for each section. Don't write directly on the drawer with a marker or anything else that's permanent. You want the flexibility of being able to change what you keep in each drawer.

Filling out the toolbox

From time to time, you may need to use ordinary workshop tools when constructing an electronics project (for example, you probably need a saw or a drill when you're building a motorized ghost display for Halloween). But don't think that you have to rush out and buy one of every tool. Depending on the type of projects that you build, you use some of these tools only occasionally. Borrow what you don't have. Just be sure to return them when you're done!

Here are some of the basics that you may want to get for yourself, or you can always figure out which neighbor has these tools and go borrowing door to door:

- ✔ **Claw hammer:** Used for just about anything you can think of that involves banging things in and prying things up. An ordinary 16-ounce claw hammer is all that you need.

- ✔ **Rubber mallet:** For gently bashing pieces together that typically resist going together. Also use a mallet for forming sheet metal, in case you're building Robby the Robot or other metal project enclosures.

- ✔ **Hacksaw:** To cut anything. Get an assortment of blades. Coarse-tooth blades cut wood and PVC pipe plastic well; fine-tooth blades work best for cutting metal.

- ✔ **Miter box:** To make angled cuts with your hacksaw. A miter box includes a three- or four-inch wide flat area where you place a board; two sides with slots in them go up on either side of the flat area forming a U-shaped channel. You put the board on the flat area, and place the saw through the slots on the sides to keep it straight while cutting the wood at an angle. Buy a good miter box and attach it to your worktable. Avoid wooden miter boxes; they don't last. An aluminum or plastic miter box is better, and only slightly more expensive.

- ✔ **Adjustable wrenches:** Sometimes called Crescent wrenches (a popular brand of wrenches), you may find these tools helpful additions to your shop.

- ✔ **Locking pliers:** Such as Vice-Grips (that's a brand name). The locking mechanism helps hold pieces for cutting, sanding, drilling, or whatever procedure that you need to do. See Figure 3-5 for an example of this tool.

- ✔ **Nut drivers:** These tools make it easy for you to attach hex nuts to machine screws. Get the assortment set because they're cheaper that way and you never know when you may need a particular size.

- ✔ **Measuring tape:** Get a cheap cloth tape at a fabric store. You don't need anything fancy or terribly long.

✔ **File assortment:** Use these files to smooth the rough edges of cut wood, metal, and plastic. Purchase a set of miniature files at the hobby store. They work just like the big-size files — they're just smaller, and therefore ideal for electronics.

✔ **Drill motor:** Get a reversible motorized drill with a variable speed control. You need to put the drill on a slow setting when working with metal and plastic. For light-duty intricate work, you can use a hand-operated drill. A drill motor with a ¼-inch or ⅜-inch chuck works fine for the electronics shop. (The *chuck* is what holds the drill bit in place. The larger the chuck, the larger the bits that can be used with the drill.)

✔ **Drill bit assortment:** You need drill bits for your drill motor. Be sure that they're sharp, and replace or sharpen them as needed. Buy an assortment of bits; the assortment that ranges from ¹⁄₃₂ inch to ¼ inch is most suitable for electronics projects.

✔ **Vise:** Used for holding parts while you work. You don't need to get elaborate — a small vise that you clamp to the edge of your worktable gets the job done.

✔ **Clear safety goggles:** Wear them when hammering, cutting, drilling, and any other time when flying debris can get in your eyes. *Be sure that you use the goggles.* Don't just display them in your workshop.

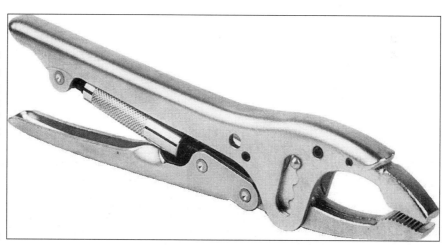

Figure 3-5:
Locking pliers work like regular pliers, but they have a locking mechanism to keep them closed in position.

Where to Park Your Tools

We talk about the basic tools that you need for electronics in the section "Oh, the Hand Tools You'll Use," earlier in this chapter. Now comes the question of where you put those tools so they're out of the way when you don't need them but readily available when you do. If you have a permanent place in your house to work on electronics, you can hang some of the tools on the wall. Reserve this special treatment for the tools that you use the most — wire cutters, pliers, and that sort of thing.

You can stash all the other tools in a small toolbox, which you can keep on your workbench. You can purchase basic toolboxes for under $10. One low-cost approach is to use a plastic fishing tackle box. (Plastic should work well because tools for electronics don't tend to be big or heavy.) A tackle box has a lot of small compartments that you can use to store screws and little parts you strip off of old projects, and the box includes a large section where you can store your basic tools.

Tools You Don't Absolutely Need (But May Find Handy)

You can use a number of tools to make your time in the electronics shop more productive and less time-consuming. These tools aren't must-haves, but if they're already out in the garage, you're sure to find a good use for them every once in a while.

Getting 'hole-istic' with a drill press

This little gadget helps you drill better holes than you can drill using a hand-held drill motor. Why? You have more control over the angle and depth of each hole. Use a drill press vice to hold the pieces — never use your hands. You find a drill press really handy when constructing your own printed circuit boards (see Chapter 11 for more about making printed circuit boards). If you outfit yourself with a small #58 drill bit, you can quickly and efficiently drill the component mounting holes in any circuit board.

You're probably most familiar with fractional size drill bits: ³⁄₃₂-inch, ⅛-inch, ¼-inch, and so on. You refer to drill bits in between the standard fractional sizes by a number value. A #58 drill bit, common in electronics because it's just the right size for making holes for component leads in printed circuit boards, is a wee 0.042 inches in diameter. The closest common fractional size is ¹⁄₆₄ inch (0.040 inch). In non-fractional terms, you call a ¹⁄₆₄-inch drill bit a #60 drill bit.

Cutting things to size with a table saw or circular saw

A table, or circular, saw is a handy item that makes cutting through large pieces of wood and plastic easier. Use a guide fence, or fashion one out of wood and clamps, to ensure a straight cut. Consult the user's guide that came with your saw if you're unsure what a guide fence or any other part of your saw is, or how to use it. Remember: safety first.

Be sure to use a fine-tooth saw blade if you're cutting through plastic. Using a saw designed for general woodcutting can cause the plastic to shatter.

Getting intricate with a motorized hobby tool

The motorized hobby tool, such as the model in Figure 3-6, may look like a small drill motor, but it spins much, much faster: 25,000 revolutions per minute and higher. (By comparison, most drill motors spin at under 2,500 revolutions per minute.) The better hobby tools, such as those made by Dremel and Weller, have adjustable speed controls.

Use the right bit for the job. For example, don't use a wood rasp bit with metal or plastic because the flutes of the rasp fill with metal and plastic debris too easily. The instructions that come with the hobby tool help you match the bits to the material that you're dealing with and the work that you're doing.

Figure 3-6:
A motor hobby tool spins at a very high speed, and you can use it to drill, cut, and shape almost any material.

Keeping Things Clean and Well-Oiled

It's a fact of life: Electronics don't like dirt. Circuitry, components, and everything you use in your electronics projects must be bright, shiny, and clean, or things may not work right. You especially should start with a clean slate if you're soldering parts to a circuit board. Dirt makes for bad solder joints; bad solder joints make for projects that either don't work at all or work only some of the time. Here are some products and techniques you can use to keep your electronics clean and tidy.

Spic-and-span electronics

You may already have most of the cleaning supplies that you need for electronics, so you may just want to make a quick run around your house to be sure that you're stocked up. Here's a checklist that you can use:

- **Soft cloth:** Keep your workshop and tools dust-free by using a soft cloth. Avoid using household dusting sprays because some generate a static charge that can damage electronics.

- **Compressed air:** You can remove dust from delicate electronic innards with a shot of compressed air. You can buy compressed air in cans, such as the one in Figure 3-7.

Figure 3-7: Canned air? You bet, and it's great for blowing dust and dirt off delicate electronics.

✔ **Household cleaner:** Lightly spray a household cleaner, such as Formula 409 or Fantastik, to remove stubborn dirt and grease from tools, work surfaces, and the exterior surfaces of your projects. Because these cleaners are water-based, don't use them around powered circuits or you may short something out.

✔ **Electronics cleaner/degreaser:** Use only a cleaner/degreaser made for use on electronic components when applying directly on parts and circuit boards. You can find the cleaner in a spray can and a bottle with a brush applicator.

Some electronics parts, especially motors, require a certain amount of grease or oil to operate. Be careful not to clean off the grease or oil that these parts need to function. If you must clean a part that requires lubrication, be sure to add fresh oil or grease when you're done.

Oil and grease to keep parts slippery

Electronics projects that use mechanical parts may require both initial lubrication and periodic re-lubrication. Case in point: a walking robot. The leg joints need a dab of oil now and then to keep things running smoothly. Whether you use oil or grease depends on what you're lubricating:

✔ For parts that spin, use a light machine oil, such as the kind you use for sewing machines or musical instruments. Avoid using oil with anti-rust ingredients because these ingredients may react to plastic parts and cause them to melt.

✔ For parts that mesh or slide, use a synthetic grease, such as lithium grease.

You can buy both light machine oil and synthetic grease at Radio Shack and other electronics stores, as well as many music, sewing machine, hobby, and hardware stores.

The Tin Man in *The Wizard of Oz* may have needed a honking big oilcan to keep himself lubed up, but in most electronics projects, a little oil goes a long way. A great alternative to oil in squeeze bottles or cans is the syringe oiler. As its name suggests, the oil is packed in a small tube that looks like a medical syringe. The "needle" is a thin, long spout, ideal for getting into hard-to-reach places. You can buy this oil at many electronics stores, as well as at some camera and music stores.

Some mechanical components don't require oil or grease, and in fact, some pieces can be harmed by lubricants. Certain self-lubricating plastics can break down if you expose them to a petroleum-based lubricant. So, unless you know for sure that a mechanical assembly or part needs oil or grease, don't automatically apply it. If you're fixing some piece of electronic gear, such as a VCR or CD player, check with the manufacturer for instructions regarding use of lubrication.

Finally, although they're convenient to use, spray-on synthetic lubricants (such as WD-40 and LPS) don't mix with electronics projects. There are two main reasons to avoid spray-on lubricants:

✔ You may have trouble controlling the coverage of the spray. The spray gets on a lot of parts that you don't want it to reach, and it makes a big mess.

✔ Many synthetic lubricants are non-conductive. The fine mist of the spray can settle on goodies that should make electrical contact with one another. If the lubricant interrupts that contact, your circuit doesn't work.

You should apply a lubricant directly and specifically to the part that needs it.

Yet more cleaning and construction supplies

You may find a variety of other cleaning, maintenance, and construction supplies handy when you're working on electronics. These supplies include

✔ **Artist brushes:** These brushes let you dust out pesky dirt. Don't get anything fancy, but avoid cheap brushes whose bristles fall out. Get both a small brush and a wide brush so that you can tackle all kinds of jobs. You can also use old toothbrushes (rinse and dry first, please).

✔ **Photographic bulb brush:** Combines the whisking action of a soft brush with the cleaning action of a strong puff of air. Get these brushes at any photo shop.

✔ **Contact cleaner:** Enables you to clean electrical contacts. The cleaner comes in a spray can, but you can apply it by spraying the cleaner onto a brush and then whisking the brush against the contacts.

✔ **Cotton swabs:** Help you soak up excess oil, lubricant, and cleaner. You can find them in quantity at any drug store.

✔ **Gauze bandage:** The larger the sheet, the better. The gauze is clean (in fact, sterile) and lint-free. You may find this material useful as a sterilized cleaning cloth for electronic parts.

✔ **Cuticle sticks and nail files:** Break out that manicure set! These personal grooming items let you scrape junk off circuit boards and electrical contacts.

✔ **Pencil eraser:** This bit of pink rubber goes a long way to rubbing electrical contacts clean, especially contacts that have been contaminated by the acid from a leaking battery. However, use care because rubbing the eraser against the circuit board may create static electricity. Be sure to use a pink eraser and not the white polymer kind. Non-pink erasers can leave a residue that can be hard to remove.

✔ **Modeling putty:** The kind of putty that you use to assemble plastic models can fill cracks and chips on the plastic exterior of your electronic projects.

Sticky Stuff to Keep Things Together

Many electronics projects require that you use adhesive of some type. For example, you may need to secure a small printed circuit board to the inside of a pocket-sized project box. A dab of glue or other adhesive does the job nicely.

Depending on the application, you can use ordinary household glue, epoxy, cyanoacrylate glue (more commonly known as super glue), double-sided foam tape, or a hot-melt glue gun. Here's the rundown on the best uses for each of these adhesives:

✔ *White household glue* is available in supermarkets, hardware stores, and home improvement stores. Household glue comes in small bottles and dries in 20 to 30 minutes (the glue takes about 12 hours to cure, however). White glue is best for projects that use wood or other porous materials. If you're using metal or plastic, opt for one of the other adhesives listed here.

✔ *Epoxy cement* comes in two tubes. To use, you mix equal parts of the tubes together and then apply the guck to the parts that you want stuck together. Most epoxies set up in five to thirty minutes and cure completely in about 12 hours. Epoxy bonds are strong and resist moisture.

✔ *Cyanoacrylate (CA) glue* bonds almost anything, almost instantly. Use it with caution because it can easily bond your fingers together. Use ordinary CA glues when bonding smooth and perfectly matching parts; use the heavier-bodied gap-filling CA glue if the parts don't mate 100 percent.

✔ *Double-sided foam tape* is a quick-and-dirty method of attaching parts. The tape works ideally in securing circuit boards to enclosures or making sure that loosely fitting components remain in place. You can cut the foam tape to almost any size that you need, and you can stack layers if you need to fill a large gap. Be sure that you get the tape and the mating surfaces dry and free of dirt before applying the tape.

✔ A *hot-melt glue gun,* such as the kind in Figure 3-8, is for the person who doesn't like to wait hours for glue to dry. Slip in a stick of glue, turn the gun on, wait a minute for it to warm up, and you can glue things with a drying time of only about 30 seconds. The glues are waterproof and gap-sealing. You apply the glue at about 250 to 350 degrees — hot enough to burn skin (so be careful!), but not hot enough to melt solder or damage most electronic components.

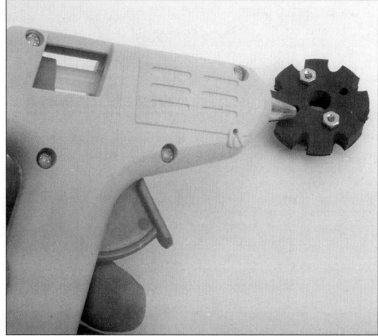

Figure 3-8:
A hot-melt glue gun quickly fastens things.

Setting Up Your Electronics Lab

Where you put your workshop is just as important as the projects you make and the tools you use. Just as in real estate, the guiding word for electronics work is location, location, location. By staking out a comfortable spot in your house or apartment, you'll be better organized and enjoy your electronics experiments much more. There's nothing worse than working with a messy workbench in dim lighting while breathing stale air.

The top ingredients for a great lab

The prime ingredients for the well-set-up electronics laboratory are

✔ A comfortable place to work, with a table and chair

✔ Good lighting

✔ Ample electrical outlets, with at least 15 amp service

✔ Tools and toolboxes on nearby shelves or racks

✔ A comfortable, dry climate

✔ A solid, flat work surface

✔ Peace and quiet

The ideal workspace won't get disturbed if you have to leave it for hours or days. Also, the worktable should be off-limits or inaccessible to young children. Curious kids and electronics don't mix!

So where in your house can you find an electronics haven, and how should you set it up to work best for your projects? The following sections give you a hand in figuring this stuff out.

Picking a perfect place to practice electronics

Before setting up shop, consider the best place in your house for building your projects. The garage is an ideal setting because it gives you the freedom to solder, hammer, and etch without having to worry about soiling the new carpet. You don't need much space; about 3 by 4 feet ought to do it. You can set aside an area for your electronics in the garage and still park the car (assuming that you don't already have that space clogged with bikes, lawn-mowers, old toys, and who knows what else).

You can use a room in the house if you don't have or can't get into your garage, but only if that room conforms to some basic requirements. When working in a carpeted room, you may want to spread another material or some protective cover over the floor to prevent static electricity — for example, use an anti-static mat. You can read more about anti-static safety measures in Chapter 2.

Putting something down to cover the floor gives you a benefit besides reducing static electricity: When the floor cover fills with solder bits and little pieces of wire and component leads, you can take the floor covering outside, beat it with a broom, and put it back as good as new. (The cover is as good as new; the broom may be a little the worse for wear.)

A bedroom, den, or family room can be an acceptable location for your electronics lab, but try to clear away a corner or section of the room for the dedicated electronics hacking. Odds are, you need to leave projects overnight or for even longer periods of time from time to time, and your work needs to stay undisturbed.

If your work area is in a main area of the house, you may want (or need) to hide it when you're not experimenting. You can make quite a mess working with electronics, especially if you're in the middle of troubleshooting your latest project. A folding screen works wonders to hide your work area, especially if you have your work surface situated in the corner of a room.

If your work area is exposed to other family members be sure to keep intergrated circuits and other sharp parts off the floor — they're painful when stepped on! Find ways to make the area off-limits to those with less knowledge about electronics safety. Kids are naturally curious about electronic gizmos, so if you have to, keep your projects, tools, and supplies out of reach on a shelf, or behind lockable doors.

If you're working in a bedroom or den, you may want to consider placing the electronics bench in the closet. Close the closet doors, and no one knows that you're building that intergalactic spaceship with built-in espresso maker.

Triple threat: Heat, cold, and humidity

No matter where you set up shop, consider the climate. If you find a work area chilly, warm, or damp, don't use that area for electronics work. Extremes in heat or cold and humidity not only make you uncomfortable; they can have a profound effect on your electronics circuits, as well.

Use these climactic clues to guide you:

- ✔ If you're working in a garage, attic, or basement, consider adding insulation if the area doesn't already have it. You can get rolls of fiberglass insulation that are relatively cheap, and installing it requires little more than a staple gun. But fiberglass can be dangerous if inhaled: Be sure to follow the installation directions carefully. For fiberglass insulation, wear gloves, eye protection, and a respirator while you're installing it.

- ✔ Some basements and garages pose a problem because they contain too much moisture. If your basement is at or below the water table level, moisture may accumulate on the floor. For safety reasons, never work in an area where the floor is wet or even slightly damp.

- ✔ When working in the garage, keep your electronics bench away from doors and other openings. This step prevents moisture from the outside from entering and ruining your projects. It also helps keep grass, bugs, and dust out of your circuit boards. (In our garage, black widow spiders like to make nests under the electronics bench. Yipes!!)

Workbench basics

You don't need a large or elaborate workbench. The types of projects that you do determine the size of the workbench, but for most applications, you need a table measuring only about 2 by 3 feet. You probably already have a small desk, table, or drafting table that you can use for your electronics bench.

Here are some other ideas for your workbench:

✔ Use a door as a table surface. Build legs using 30-inch lengths of 2-by-4 lumber and attach the legs using joist hangars. (You can buy all this stuff at any home improvement store.) You can get hollow-core doors for less money, but the solid-core doors last longer and don't bow with age and weight. As an alternative, you can build your work surface using ¾-inch plywood or particle board.

✔ If you prefer, forget the 2-by-4 legs and make a simple table surface using a door and two sawhorses. The advantage of this get-up is that you can take apart and store your workbench in the corner when you're not using it.

✔ Many electronics technicians prefer to cover their work surface with a layer of carpeting. The carpeting acts as a cushion to protect circuit boards, cabinets, and other components. If you use a piece of carpet, get a new, clean remnant and cut it to size. The shorter the nap, the better (so that you don't constantly lose little parts in the shag). If you can, get a carpet that has been treated with an anti-static spray or, better yet, contains anti-static metallic threads.

Remember, as you work on projects, you crouch over the worktable for hours at a time. You can skimp and buy or build an inexpensive worktable, but if you don't already own a good chair, put one on the top of your shopping list. Be sure to adjust the seat for the height of the worktable. A poor-fitting chair can cause backaches and fatigue.

Chapter 4

Getting to Know You: The Most Common Electronic Components

In This Chapter

▶ Getting the lowdown on resistors

▶ Quickly changing resistance with potentiometers (and why you'd do this)

▶ Discovering how to pick the best capacitor for your circuit

▶ Decoding common markings on resistors and capacitors

▶ Delving into diodes, including the kind that light up

▶ The truth about transistors

▶ Understanding integrated circuits

*E*lectronics folks refer to the assortment of odds-and-ends that go into a circuit, collectively, as components. These are the things that make a circuit work. Although you can make a complete circuit with just a battery, some wire, and a light bulb, most electronics projects use a few more components, such as resistors, capacitors, diodes, transistors, and integrated circuits. You can think of these components as the common building blocks of the typical electronics gizmo.

The variety of components, and the way they connect to one another, determines what a circuit does. Wired one way, a collection of a few resistors, capacitors, and transistors can build an electronic siren; wired another way, the circuit can become a flashing crossing sign for a model railroad.

In this chapter, you can read about the most common electronic components used in circuits: what they are, what you can use them for, and what they do. And because part of becoming an electronics pro includes being able to identify components just by the way they look, you discover how to do that in this chapter, as well.

Viva la Resistors

Electric current is simply the motion of electrons from one place to another through a wire. The more electrons that are flowing, the higher the current. *Resistors* have an apt name: They "resist" the electrical current going through them. You can think of resistors as "brakes" for electrons. By controlling the electrons going through a resistor, you can make a circuit do different things.

Resistors may be the primary building block of circuits, so you see them quite a bit in electronics projects. Here are some of the things you can use them for:

- ✔ **Limiting current to another component:** Some parts, such as light emitting diodes (LEDs), eat up current. Like a kid eats candy bars they try to gobble up as much as you give them. But LEDs run into a problem — they burn themselves out if they eat too much current. You can use a resistor to limit the amount of current that reaches an LED.

- ✔ **Reducing voltage to part of the circuit:** In many circuits, you need to provide different voltages to different parts of the circuit. You can do this easily with resistors. Two resistors joined, as Figure 4-1 shows you, form what's called a voltage divider. Assuming that you have two identical resistors, that is, they apply their brakes in the same amount, the voltage in between the two is exactly half that of the rest of the circuit.

- ✔ **Controlling the voltage/current going into another component:** Combine a resistor and a capacitor, for example, and you create a kind of hourglass timer. Or put a resistor at the input of a transistor to control how much the transistor amplifies a signal. Or . . . well, you get the idea.

Figure 4-1:
Use two resistors to create this voltage divider, a common technique to produce different voltages for different parts of a circuit.

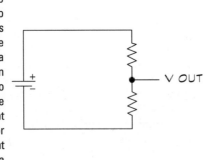

✔ **Protecting the inputs of sensitive components:** Too much current destroys electronic components. By putting a resistor at the input of a sensitive transistor or integrated circuit, you limit the current that reaches that transistor or circuit. Although not foolproof, this simple technique can save you a lot of time and money that you would lose fixing accidental blow-ups of your circuits.

Ohming in on resistor values

If resistors act like brakes, then you have to have some way to change how hard you push the pedal, in order to have control over the flow of electrons. That control involves modifying the resistance of a resistor.

Electronics dabblers know the amount of resistance in a resistor as the ohm, typically represented by the Greek capitalized letter omega: Ω. The higher the ohm value, the more resistance the component provides.

To understand how you can adjust resistance, you should know that there are two basic types of resistors, fixed and variable. Here's how they differ:

✔ A *fixed resistor* supplies a pre-determined resistance to current. Color coding identifies the value of most fixed resistors. The color coding starts near the edge of the resistor and is comprised of four, five, and sometimes six bands of different colors. Figure 4-2 shows the order of bands marked on the body of the resistor along with what each represents.

✔ A *variable resistor,* called a potentiometer, allows for the continual adjustment from virtually no ohms to some maximum value. The potentiometer usually has the maximum value printed on it somewhere. See the section "Dialing with potentiometers," later in this chapter, for detailed info on these puppies.

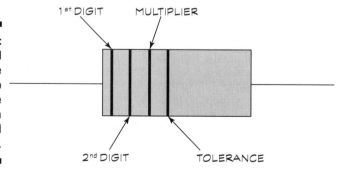

Figure 4-2:
Color coded bands are used to denote the value in a fixed resistor.

Not all resistors use color coding. Sometimes, the exact value may be printed on the resistor. This is typical of so-called *precision resistors:* The actual resistance of the component is very close to or exactly what you see printed on them. You can read more about the precision of resistors in the following section.

Color me red, green, and blue

As we noted in the previous section, the vast majority of resistors use color coding to tell you what resistance, in ohms, they provide. The color code is a world-wide standard, and we've been using it in electronics for many decades. Although the colors are standardized, a resistor can have either four or five bands of color, depending on whether it's standard-precision or high-precision.

Standard-precision resistors use four color bands. These resistors come within at least 2 percent of their marked value. That is, the markings on the resistor and the actual value of the resistor when you test it fall within at least 2 percent of one another. You use standard-precision resistors for 99 percent of your hobby projects. *High-precision resistors* have five color bands, and they come within 1 percent or less of their marked value. You can find out more about high-precision resistors in the section "A word (or two) about high-precision resistors," later in this chapter.

Here's what the bands on a standard-precision resistor represent:

✔ Bands one, two, and three indicate the value of the resistor.

✔ Band four indicates the tolerance of the resistor and typically falls within +5 percent or +10 percent of the resistor's actual tolerance (a range of resistance value; read more about this in the following section).

Table 4-1 shows the meaning of the color codes used on the bands so that you can determine the value of the resistor. Assume that the resistor has yellow-violet-red-silver markings. The first two bands indicate the first two digits of the value of the resistor. Referring to Table 4-1, yellow represents 4 and violet stands for 7, so the significant digits of a resistor with a yellow-violet-red-silver band scheme are 47. The third band indicates the multiplier, in this example that band is red, so the value is 100. Multiply 47 by 100, and you get 4700 ohms. You express values over 1000 in K (for kilo, or 1000), so you say that the resistor has a value of 4.7K ohm. Note that in this table certain colors will never be used for certain bands, hence no value is noted.

Table 4-1		Resistor Color Coding		
Color	*1st Digit*	*2nd Digit*	*Multiplier*	*Tolerance*
Black	0	0	1	+20%
Brown	1	1	10	+1%
Red	2	2	100	+2%
Orange	3	3	1,000	+3%
Yellow	4	4	10,000	+4%
Green	5	5	100,000	n/a
Blue	6	6	1,000,000	n/a
Violet	7	7	10,000,000	n/a
Gray	8	8	100,000,000	n/a
White	9	9	n/a	n/a
Gold			0.1	+5%
Silver			0.01	+10%

Understanding resistor tolerance

The last band of the resistor indicates its tolerance. Tolerance takes into account unavoidable variations in resistor manufacturing. Though a resistor may have a 2,000 ohms marking, for example, its actual value may be slightly higher or lower. You refer to the potential variation in value as *tolerance,* expressed as a percentage (for example, ±5-percent tolerance means the resistor value may vary plus or minus 5 percent from the stated value). In most cases, being a little off doesn't significantly affect the operation of the circuit. Knowing the tolerance of the resistor lets you decide if a resistor is adequate for a particular circuit. Tolerance appears in the last column of Table 4-1.

Take a look at the yellow-violet-red-silver resistor example from the previous section. By looking at Table 4-1, you can see that silver, the last band, denotes +10-percent tolerance. This value means that the resistor can vary in tolerance 10 percent higher or lower than its indicated value. If the resistor has an indicated value of 4.7K with a 10 percent tolerance, the actual value can fall anywhere between 4,230 and 5,170 ohms.

A word (or two) about high-precision resistors

Many high-precision resistors use a five-band color coding system (on those that don't, the actual value of the resistor is printed on its body). These resistors have a tighter tolerance than standard-precision resistors. You use high-precision resistors in a circuit where you need to have a resistor of a very specific value. For example, a resistor used in a timing or voltage reference circuit may need such a precise value.

In a high-precision resistor, here's what the bands represent:

✔ Bands one through four indicate the value of the resistor.

✔ The fifth band indicates the tolerance of the resistor, typically +1 percent.

Most circuits tell you the safe resistor tolerance to use, either for all the resistors in the circuit or for each resistor. Look for a notation in the parts list or as a footnote at the bottom of the circuit diagram. If the schematic doesn't state a tolerance, then you can assume that you may safely use standard +5 percent or +10 percent tolerance resistors.

If you aren't sure of the resistance of a particular resistor, you can use a multimeter to check it, as we describe in Chapter 9.

Let there be heat

Whenever electrons flow through something, they generate heat. The more electrons, the higher the heat. Resistors also come rated by their power. Power is measured in watts — the higher the watts, the higher the heat. Electronic components can only stand so much heat (how much depends on the size and type of component) before they sizzle into a charred mass. The power rating tells you how many watts can safely go through the resistor. You calculate watts by using this formula:

$$P = I \times V$$

P stands for power, measured in watts; I represents the current, in amps, going through the resistor; and V represents the voltage as measured across the resistor. For example, suppose that the voltage is 5 volts, and 25 milliamps of current go through the resistor. To calculate watts, multiply 5 by 0.025. You get 0.125, or ⅛ watt.

Unlike the value in ohms, the component seldom has the resistor wattage printed on it, either written out or as part of the color code. Instead, you have to figure out the wattage by the size of resistor, or, if you know where you bought the resistor, check with the manufacturer. Resistors that you use in high-load applications, such as motor or lamp control, require higher wattages than those resistors that you use in low-current applications. The majority of resistors that you use for hobby electronics are rated at ¼ or even ⅛ watt.

High-wattage resistors take many forms, some of which you can see in Figure 4-3. Resistors over five watts commonly come encased in epoxy or other waterproof and flameproof coating and have a rectangular, rather than cylindrical, shape. Higher-wattage resistors may even include their own metal heat sink where the fins help draw heat away from the resistor.

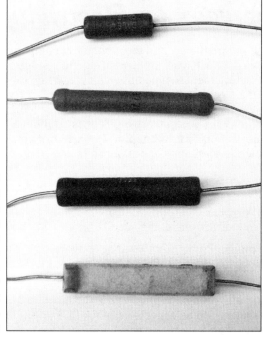

Figure 4-3: High-wattage resistors come packaged in many different forms.

Dialing with potentiometers

Variable resistors, more commonly known as potentiometers (or in electronics slang as pots), let you "dial in" a resistance. The upward value of the potentiometer determines the actual range of resistance. Most potentiometers are marked with this upward value — 10K, 50K, 100K, 1M, and so forth. For example, with a 50K potentiometer, you can dial in any resistance from 0 to 50,000 ohms. Bear in mind that the range on the potentiometer is approximate only. If the potentiometer lacks markings, you need to use a multimeter to figure out the component's value. (You can read about how to test resistances, using a multimeter, in Chapter 9.)

Virtually all potentiometers are the dial type. You sometimes use the other type, the slide potentiometer, on gadgets like stereo equipment. The dial kind is typically easier to mount into your own projects.

With the dial type of potentiometer, you can rotate the dial nearly 360 degrees, depending on the specific qualities of the potentiometer that you're using. At one extreme, the potentiometer has zero resistance going through it; at the other extreme, the resistance is the maximum value of the component. Your television volume control or electric blanket control are typical examples of the dial pot.

Capacitors: Reservoirs for Electricity

After resistors, capacitors are the second most common component in the average electronic device. Capacitors are interesting little gadgets. They store electrons by attracting them to a positive voltage. When the voltage is reduced or removed, the electrons move off. When a capacitor removes or adds electrons to the circuit in this fashion, it can work to smooth out voltage fluctuations. In some cases you can use capacitors combined with resistors as timers (read more about this in Chapter 7). Capacitors make possible all kinds of circuits, such as amplifiers and thousands of others.

Capacitors are used for all sorts of neat applications, including

- **Creating timers:** A kind of electronic metronome, a timer most often pairs up with a resistor to control the speed of the tick-tick-tick.

- **Smoothing out voltage:** Power supplies that convert AC current to DC often use capacitors to help smooth out the voltage so that the voltage stays at a nice, constant level.

- **Blocking DC current:** When connected inline (in series) with a signal source, such as a microphone, capacitors block DC current but pass AC current. Most kinds of amplifiers use this function, for example.

- **Adjusting frequency:** You use capacitors to make simple filters that reject AC signals above or below some desired frequency. By adjusting the value of the capacitor, it's possible for you to change the cut-off frequencies of the filter.

A quick look inside a capacitor

Though they many sound complicated because of all the things that you can use them for, capacitors are really very simple devices. The typical capacitor has two metal plates inside it. The plates don't touch. Instead, a *dielectric material,* which is a fancy term for an insulator, separates the plates.

Common dielectrics used in capacitors include plastic, mica, and paper. (We talk more about the dielectric in the section "Dielectric this, dielectric that," later in this chapter.)

Farads big and small

You probably realize by now that, just as politicians have an excuse for everything, electronics types have units of measure for absolutely everything. The electronics world rates capacitors by capacitance, expressed in *farads*. The higher the value, the more electrons the capacitor can store at any one time. The farad is a rather large unit of measurement, so the bulk of capacitors available today come rated in microfarads, or a millionth of a farad. You may even come across an even smaller rating — the picofarad, or a millionth of a millionth of a microfarad. Using the Greek "micro" character, electronics documentation most often shortens the microfarad to µF, as in 10µF. You shorten the picofarad to the simple pF.

Here are some examples:

- ✔ A 10-µF capacitor is 10 millionths of a farad.
- ✔ A 1-µF capacitor is 1 millionth of a farad.
- ✔ A 100-pF capacitor is 100 million of a millionth of a microfarad.

Keeping an eye on the working voltage

The *working voltage,* sometimes abbreviated simply as WV, is the highest voltage that a capacitor can withstand before the dielectric layers in the component become damaged. At higher voltages, the current may simply arc between the plates, like a lightning strike during a storm. If a capacitor isn't designed to withstand high voltages, a spark develops within the capacitor that punches through the dielectric material, leaving the component useless (shorted out).

The typical capacitor designed for DC circuits rates at no more than 16 to 35 volts. You don't need higher voltages because anywhere between 3.3 and 12 volts typically powers these circuits. Only when you build circuits that use higher voltages do you need to concern yourself with the working voltage of capacitors. It's a good idea to select a capacitor with a working voltage of at least 10-15 percent more than the voltage in the circuit for safety.

Dielectric this, dielectric that

Suppose your teen asks you to make her a banana split. The problem: You don't have any bananas. So you improvise and use cucumbers instead. Blecch!

It's just not the same. It's not the cuke's fault, it's how you used the hapless vegetable that spelled culinary disaster. Similarly, designers of electronic circuitry specify capacitors for projects by the dielectric material in them. Some materials are better in certain applications: Just like bananas in a banana split, they provide a better match.

The most common dielectric materials are aluminum electrolytic, tantalum electrolytic, ceramic, mica, polypropylene, polyester (or Mylar®), paper, and polystyrene. If a circuit diagram calls for a capacitor of a certain type, you should be sure to get one that matches.

Table 4-2 lists the most common capacitor types, their typical value range, and common applications.

Table 4-2		Capacitor Characteristics
Type	**Range**	**Application**
Ceramic	1 pF to 2.2 µF	Filtering, bypass
Mica	1 pF to 1 µF	Timing, oscillator, precision circuits
Metalized foil	to 100 µF	DC blocking, power supply, polycarbonate filtering
Polyester	.001 to 100 µF	Same for polycarbonate
Polystyrene	10pF to 10 µF	Timing, tuning circuits
Paper foil	.001 to 100 µF	General purpose
Tantalum	.001 to 1000 µF	Bypass, coupling, DC blocking
Aluminum	10 to 220,000 µF	Filtering coupling, bypass electrolytic

Big capacitor in itty-bitty living space

Making farad-range capacitors has become possible only recently. Using older construction techniques, a one-farad capacitor would be bigger than a bread box and kind of unwieldy.

By using other technologies and materials, such as microscopically-small carbon granules, manufacturers can now build capacitors of one farad and above that fit into the palm of your hand. Computer memories, clock radios, and other electric devices that need to retain a small charge for extended periods of time when they have no access to power routinely use capacitors as substitute batteries.

Capacitors come in a variety of shapes, as you can see in Figure 4-4. Aluminum electrolytic and paper capacitors commonly come in a cylindrical shape. Tantalum electrolytic, ceramic, mica, and polystyrene capacitors have a more bulbous shape because they typically get dipped into an epoxy or plastic bath to form their outside skin. However, not all capacitors of any particular type (such as mica or Mylar®) get manufactured the same way, so you can't always tell the component book by its cover.

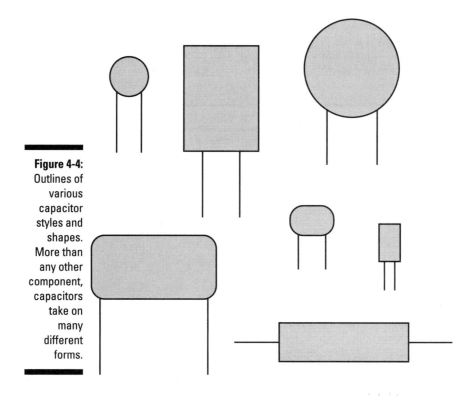

Figure 4-4: Outlines of various capacitor styles and shapes. More than any other component, capacitors take on many different forms.

How much capacity does my capacitor have?

Some capacitors have their value in farads or portions of a farad printed directly on them. You usually find this to be the case with larger aluminum electrolytic types; the large size of the capacitor provides ample room to print the capacitance and working voltage.

Most smaller capacitors, such as 0.1- or 0.01-µF mica disc capacitors, use a three-digit marking system to indicate capacitance and tolerance. Most folks find the numbering system easy to use. But there's a catch! (There's always a catch.) The system is based on picofarads, not microfarads. A number using this marking system, such as 103, means 10 followed by three zeros, as in 10,000 or 10,000 picofarads.

Any value over 1,000 picofarads most often comes in microfarads. To make the conversion from picofarads to microfarads, just move the decimal point to the left six spaces. So, the result of the example above (with its 10,000 pF) is 0.01 µF.

You can use Table 4-3 as a handy reference guide to common capacitor markings that use this numbering system. Notice that two-digit values are in picofarads. So, for instance, a capacitor with "22" printed on it has a value of 22 picofarads. Three digit numbers are microfarads.

Table 4-3	Capacitor Value Reference
Marking	*Value*
nn (a number from 01 to 99)	nn pF
101	0.0001 µF
102	0.001 µF
103	0.01 µF
104	0.1 µF
221	0.00022 µF
222	0.0022 µF
223	0.022 µF
224	0.22 µF
331	0.00033 µF
332	0.0033 µF
333	0.033 µF
334	0.33 µF
471	0.00047 µF
472	0.0047 µF
473	0.047 µF
474	0.47 µF

Another, less-often-used numbering system uses both numbers and letters, like this:

4R1

The placement of the letter R tells you the position of the decimal point: 4R1 is really 4.1. This numbering system doesn't indicate the units of measure, however, which can be in microfarads or picofarads.

You can test capacitance with a capacitor meter or a multimeter with a capacitance input. Most meters require that you plug the capacitor directly into the test instrument, as the capacitance can increase with longer leads. This makes the reading less accurate. Chapter 9 talks about testing capacitors.

When a microfarad isn't quite a microfarad

Most capacitors are rather inexact beasts. The value printed on the capacitor, and the actual capacitance of the capacitor, may not be the same. In fact, they may not even come close. Manufacturing variations cause this problem; capacitor makers aren't just out to confuse you. Fortunately, the inexactness is seldom an issue in homebrewed circuits. Still, you need to know about these variations so that, if a circuit calls for a higher precision capacitor, you know what to buy.

Like resistors, capacitors are rated by their tolerance, and this tolerance comes as a percentage. On many capacitors, a single letter code indicates the tolerance. You may find that letter placed by itself on the body of the capacitor or placed after the three digit mark, such as

103Z

The letter Z denotes a tolerance of +80 percent to -20 percent. This tolerance means that the capacitor, rated at 0.01 µF, may have an actual value as much as 80 percent higher or 20 percent lower than the stated value. Table 4-4 lists the meanings of common code letters used to indicate capacitor tolerance.

Table 4-4	Capacitor Tolerance Markings
Code	*Tolerance*
B	+ 0.1 pF
C	+ 0.25 pF

(continued)

Table 4-4 (continued)

Code	Tolerance
D	+ 0.5 pF
F	+ 1%
G	+ 2%
J	+ 5%
K	+ 10%
M	+ 20%
Z	+ 80%, -20%

Tolerating hot and cold

Here's a little tolerance gotcha: The value of a capacitor changes with temperature, which you call its *temperature coefficient.* When the markings on a capacitor mention temperature coefficient at all, the value is indicated as a three-character code, such as NP0, meaning negative/positive zero. A capacitor with the NP0 designation is fairly tolerant of temperature changes.

More and more capacitor manufacturers are adopting what they call the EIA marking system for temperature tolerance, which you can check out in Table 4-5. The three characters in each mark indicate the temperature tolerance and maximum variation within the stated temperature range.

For example, using Table 4-5, you can figure out that a capacitor marked Y5P has the following characteristics:

- ✔ -30+C low temperature requirement
- ✔ +85+C high temperature requirement
- ✔ +10.0 percent variance in capacitance within the -30+C to +85+C range

Table 4-5 **EIA Capacitor Codes**

1st Letter Symbol	Low Temp. Requirement	Number Symbol	High Temp. Requirement	2nd Letter Symbol	Max. Capacitance Change Over Temperature Rating
Z	+10 C	2	+45 C	A	+1.0%
Y	-30 C	4	+65 C	B	+1.5%

1st Letter Symbol	Low Temp. Requirement	Number Symbol	High Temp. Requirement	2nd Letter Symbol	Max. Capacitance Change Over Temperature Rating
X	-55+ C	5	+85 C	C	+2.2%
		6	+105 C	D	+3.3%
		7	+125 C	E	+4.7%
				F	+7.5%
				P	+10.0%
				R	+15.0%
				S	+22.0%
				T	+22%, −33%
				U	+22%, −56%
				V	+22%, −82%

Being positive about capacitor polarity

One final mark that you find on some capacitors, especially tantalum and aluminum electrolytic types, is a polarity symbol. By convention, most capacitors use the minus (–) sign for the negative terminal and don't use the plus (+) sign for the positive terminal. For example, as the top capacitor in Figure 4-5 shows, the minus sign and arrow point to the negative lead of the aluminum electrolytic capacitor.

Note that only larger-value capacitors (1 µF and up), typically just the electrolytic types, are polarized. (You can still find non-polarized electrolytic capacitors out there, too. They're commonly used in stereo speaker systems.) The smaller capacitors, such as mica, ceramic, and Mylar®, are not polarized, so they don't have a polarity mark.

If a capacitor is polarized, you *really, really* need to make sure to install it in the circuit with the proper orientation. If you reverse the leads to the capacitor, by connecting the + side to the ground rail, for example, you may ruin the capacitor. You may also damage other components in the circuit, or the capacitor could even explode.

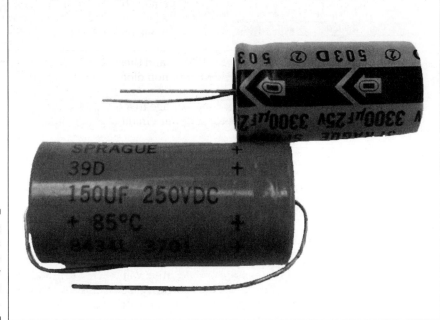

Changing capacitance

It's always a good thing when you get things just the way you want them. That's why it's so nice that variable capacitors allow you to adjust capacitance to suit your needs.

The most common type of variable capacitor that you encounter is the air dielectric type, such as the one you find in the tuning control of an AM radio. Smaller-variable capacitors are often used in radio receivers and transmitters, and they work in circuits that use quartz crystals to provide an accurate reference signal. The value of such variable capacitors typically falls in the 5 to 500 pF range.

Diode Mania

The *diode* is the simplest form of semiconductor. You use semiconductors in a circuit to control the flow of electrons (Chapter 1 tells you more about semiconductors). A diode has two terminals, each with a high resistance to current in one direction and low resistance to current in the opposite direction. Or put another way, diodes act as a one-way valve for electrons. Electrons can go through the diode in one direction but not in the other.

A variety of applications use diodes, and these diodes fall into numerous sub-types. Here is a list of the most common diodes:

✔ **Zener:** These puppies limit voltage to a pre-determined amount. You can build a voltage regulator for your circuit cheaply and easily with a zener diode.

✔ **Light-emitting diode (LED):** All semiconductors emit infrared light when they conduct current. LEDs emit visible light. Now available in all the colors of the rainbow.

✔ **Silicon-controlled rectifier (SCR):** The SCR is a type of switch used to control AC or DC currents. They're common in light dimmer switches.

✔ **Rectifier:** This basic diode transforms (referred to as "rectifying") AC current to provide DC current only. (***Remember:*** AC current alternates between both positive and negative values. DC current does not alternate, and is only positive or negative. See Figure 4-6 for an example.) Diodes are often referred to as rectifiers because they perform this rectifying function.

✔ **Bridge rectifier:** This component consists of four diodes, connected one to the other to form a kind of box shape; it rectifies AC to DC with maximum efficiency.

You can see a sampling of diodes in Figure 4-7.

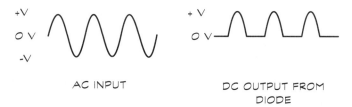

AC INPUT

DC OUTPUT FROM DIODE

Figure 4-6: Diodes can transform AC current to DC current.

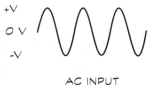

AC INPUT

DC OUTPUT FROM BRIDGE RECTIFIER

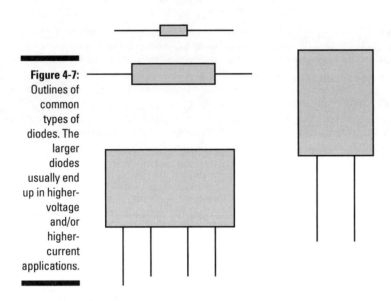

Figure 4-7:
Outlines of
common
types of
diodes. The
larger
diodes
usually end
up in higher-
voltage
and/or
higher-
current
applications.

Important ratings for diodes: Peak voltage and current

Except for zeners, diodes don't have "values" like resistors and capacitors. A diode simply does its thing in controlling the flow of electrons. But that doesn't mean all diodes are the same. Diodes are rated by two main criteria: peak inverse voltage (PIV) and current. These criteria specify the kind of diode that you should use in a given circuit.

- The *PIV rating* roughly indicates the maximum working voltage for the diode. For example, if the diode is rated at 100 volts, you shouldn't use it in a circuit that applies more than 100 volts to the diode.

- The *current rating* is the maximum amount of current the diode can withstand. Assuming a diode is rated for 3 amps, it can't safely conduct more than 3 amps without overheating and failing.

Diodes are identified by an industry-standard numerical system. A classic example is the 1N4001 rectifier diode, which is rated at 1.0 PIV and 50 volts. A 1N4002 is rated at 100 volts, a 1N4003 is rated at 200 volts, and so on. We promise not to bore you with what all the numbers represent and how they correspond to PIV and current: You can readily find this information in any diode data cross-reference book or electronics component catalog.

Want to become a diode spotter? Rectifier diodes rated to about 3 to 5 amps generally come encased in black or gray epoxy, and they're designed so that you can directly mount them on printed circuit boards. Higher-current diodes, such as 20, 30, or 40 amps, commonly come contained inside a metal housing. The metal housing includes a heat sink or a mounting stud so that you can affix the diode securely on a heat sink. A few diodes use the same packaging as transistors (which we describe in the next section).

Which way is up?

All diodes have what amounts to positive and negative terminals. The terminals go by special names: The positive terminal is called the *anode,* and the negative terminal is called the *cathode.* You can readily identify the cathode end of a diode by looking for a red or black stripe near one of the leads. Figure 4-8 shows a diode with a stripe at the cathode end. This stripe corresponds with the line in the schematic symbol for the diode. It's important that when you follow a schematic to build a circuit you orient the diode with the line facing the specified way.

Figure 4-8:
Keep polarity in mind when using diodes. The stripe on a diode marks its cathode.

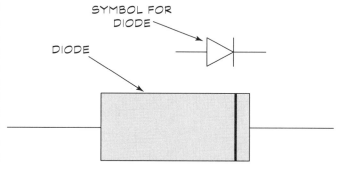

As we talk about in the section "Diode Mania," earlier in this chapter, diodes pass current going in one direction and block current going in the other. So, if you insert a diode backward in a circuit, either the circuit doesn't work at all or you damage some components. Always note the orientation of the diode when you use it in a circuit. Double-check to make sure that you have it right!

Fun, fun, fun with light-emitting diodes

If bright lights turn you on, you can appreciate the curious behavior of semiconductors: They emit light when you apply an electric current to them. This light is generally very dim and only in the infrared region of the electromagnetic spectrum. The light-emitting diode (LED), such as the light that glows yellow or green when your computer is on, is a special type of semiconductor expressly designed to emit copious amounts of light. Most LEDs are engineered to produce red, yellow, or green visible light, but some special-purpose types emit infrared, blue, and even white light.

LEDs carry the same specifications as any other diode, but they usually have a pretty low current rating. An LED has a PIV rating of about 100 to 150 volts, with a maximum current rating of under 50 milliamps. If more current passes through an LED than its maximum rating allows, the LED burns up like a marshmallow in a campfire.

LED specifications indicate both the maximum current rating, usually referred to as *forward current,* and the peak current. The *peak current* is the absolute maximum current that you can pass through the LED for a very short period of time. Here, short means short — on the order of milliseconds. Don't confuse forward current with peak current, or you may wreck your LED.

Resistors, meet LEDs

You use a resistor, such as the one in Figure 4-9, to limit the current to the LED. You select the value of the resistor to maintain the current below the maximum current rating of the LED. The calculation is simple, and for most LEDs and 5 or 12 volt circuits, you can use common resistor values that get you in the right ballpark.

Figure 4-9:
A resistor inserted in series with an LED is used to limit current to the LED.

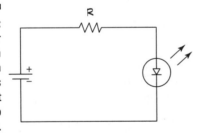

We list common resistor values in Table 4-6; the values are selected based on the ratings of most LEDs.

Table 4-6	Resistor Values Used with LEDs
Circuit Voltage	*Current Limiting Resistor Value*
3.3 to 5 volts	330 ohms
6 to 9 volts	560 ohms
12 to 15 volts	1K ohms

You can always select a higher value resistor, which simply makes the LED glow less brightly. If you select a lower value resistor, you run the risk of burning out the LED. Because most LEDs cost just a few pennies a piece, you can experiment with resistors of different values and not break the bank. Make it a game to see how bright you can make your LED before you make it go up in smoke — just kidding!

If you want a more accurate calculation, you need to know the forward voltage drop through the LED, in addition to the LED's maximum current rating. Most standard brightness LEDs have a forward voltage drop of about 1.5 volts. The latest crop of ultra-bright LEDs may have forward voltage drops exceeding 3.5 volts.

The calculation for desired forward current, in egghead terms, is this:

$R = (V_s - V_f) / I_f$

- ✔ **R** stands for the value of the resistor, in ohms, that you want to use.

- ✔ **V_s** represents the supply voltage. It's measured in volts.

- ✔ **V_f** is the forward voltage drop through the LED. This is also measured in volts

- ✔ **I_f** stands for the forward current (in amps) that you want to pass through the LED. You can use the maximum current rating of the LED or something less for the forward current, but never use more!

Suppose a circuit is powered at 6 VDC and the forward voltage drop through the LED is 1.2 volts. You want a forward current of 40 mA (that's 0.040 of an amp). Substituting these values in the calculation, you get:

$R = (6 - 1.2) / 0.040$

Do the math in your head, on paper, or with a calculator, and you see that R equals 120 (ohms). So, to pass 40 mA of current through this particular LED when using a 6-volt supply, you use a 120-ohm resistor. Remember to do the calculation again if you change the voltage of the power supply or use an LED with a higher or lower forward voltage drop.

The Transistor: A Modern Marvel

Imagine the world without the simple transistor. Radios would all be the size of a toaster oven. Cell phones would be the size of a washing machine. And today's super fast computers would be the size of . . . Rhode Island.

Transistors were developed as an alternative to the vacuum tube. The two main ways that you can use transistors (or vacuum tubes, for that matter) are to amplify a signal or to switch a signal on and off. Besides its small size, a transistor has another advantage — it uses less power than a vacuum tube to accomplish the same job.

With creative connections in a circuit, you can also use transistors to switch or amplify voltages. This fancy circuit work can confuse you when you're studying circuits involving transistors. Transistors are very complex little critters, so we just talk about the basic types you encounter when you begin working in the electronics world, what they look like, and other getting-to-know-you details in this book.

Millions of individual transistors make up the microprocessor at the heart of your home computer. Without transistors, we would live in a world with no PCs. (Hmmm...late at night, slaving away on my computer, I think maybe transistors aren't such a great idea after all...)

Slogging through transistor ratings

Resistors, capacitors, and even diodes have fairly simple and straightforward ratings. But the transistor just has to be difficult. These doohickeys are rated by a number of criteria far too extensive for this book to tackle. Here are just a very few of them:

- ✔ Collector-to-base voltage
- ✔ Collector-to-emitter voltage
- ✔ Maximum collector current
- ✔ Maximum device dissipation
- ✔ Maximum operating frequency

I am not a number, I'm a free transistor!

At last count, you can find several *thousand* different transistors currently available from more than two dozen manufacturers. How can you tell them apart? A unique number code, such as 2N2222 or MPS6519, identifies each kind of transistor. If you're rebuilding a circuit that you see in a book or on a Web page, use the transistor number code to find a match.

If you can't find an exact match, you can probably use a close substitute. Transistor manufacturers provide substitution guides that help you find which one of their parts closely matches the transistor that you're looking for. NTE, a major reseller of replacement transistors, provides a popular transistor substitution guide. Visit their Web site at www.nteinc.com for an online cross-reference substitution guide.

None of these ratings appears anywhere on the transistor — that would be too easy. To determine these characteristics, you have to look up the transistor in a specifications book, or consult the technical documentation at the manufacturer's Web site. For basic electronics tinkering, you don't need to know — or even understand — what these specifications mean. More than likely, you simply use the transistor that your project specifies.

Figure 4-10 shows a grab bag of different kinds of transistors.

On the case of transistor cases

The semiconductor material in a transistor is the size of a grain of sand or even smaller. It's kind of hard to solder wires to something so teensy, so they put semiconductor material in a metal or plastic case. You can find literally dozens and dozens of sizes and styles of transistor cases, and this book definitely can't describe them all. But to help you identify the most common types, here's what you should look for:

✔ **Plastic or metal:** Signal transistors come in either plastic or metal cases. The plastic variety works for most uses, but some precision applications need the metal variety because transistors that use metal cases (or *cans*) are less susceptible to stray radio frequency interference. Signal transistors almost always have three lead connections (sometimes four). If the transistor has just two wires, it's probably the light-dependent type, which we talk about in Chapter 5.

✔ **Size matters:** Power transistors come in both plastic and metal cases, and they're physically larger than signal transistors.

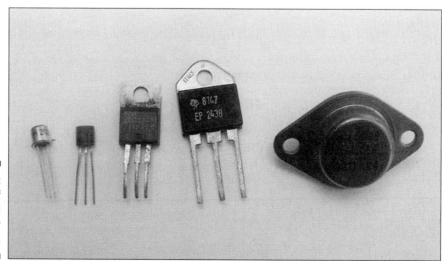

Figure 4-10:
A sampler of signal and power transistors.

Making connections

Transistors typically have three wire leads. The leads in the typical three-lead transistor are

✔ Base

✔ Emitter

✔ Collector

A base is wired to a voltage or current and turns the transistor on or off. Emitter and collectors leads connect to a positive or negative voltage source or ground. Which lead goes where varies with the circuit.

You can see this arrangement of connections in Figure 4-11. A few transistors, most notably the field-effect transistor (or FET), include a fourth lead. This lead grounds the case to the chassis of the circuit.

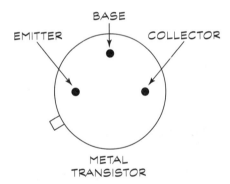

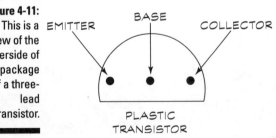

Figure 4-11: This is a view of the underside of a package of a three-lead transistor.

You absolutely, positively have to make sure that you don't install a transistor the wrong way in your circuit. Switching the connections around can damage the transistor and sometimes other components. You can find yourself even more confused by transistor connections because they're often (though not always!) shown from the underside of the case because that's where you solder them on. That is, the lead pinouts look as if you've turned the transistor over and are looking at it from the bottom. This perspective makes soldering the transistor in a circuit board easier.

Transistor types

First, transistors qualify as either NPN or PNP devices. These mysterious abbreviations refer to the sandwiching, or junctions, of the semiconductor materials inside the device. Unless you have x-ray vision, you can't tell the difference between an NPN and PNP transistor just by looking at them. However, the catalog specification sheets, as well as schematics, should tell you the difference, as in Figure 4-12. You select NPN and PNP devices based on how you plan to use the transistor in the circuit. We can't get into the nitty-gritty of choosing an NPN or PNP transistor here because it could fill a whole book. But we can say that you can't mix-and-match NPN and PNP transistors. If a circuit calls for a PNP transistor, you can't substitute an NPN type without expecting to see smoke billowing out of some part of your device.

Figure 4-12:
Schematic symbols for NPN and PNP transistor types. For an NPN, the arrow points out from the center of the symbol. For a PNP, the arrow points in.

NPN TRANSISTOR   PNP TRANSISTOR

As if you didn't have enough stuff to memorize, in addition to the junction type, transistors are categorized by how the junction is created during manufacturing. The two main types of transistors that you're likely to encounter are bipolar and FET. Here's how they differ:

✔ **Bipolar transistors:** These transistors are the most common kind (and they're the kind that we cover in the preceding sections). A small input current is applied to the base of the transistor. This in turn, changes the amount of current that flows between the collector and emitter.

✔ **FETs (field effect transistors):** These transistors also have three connections, but you call those connections gate, source, and drain, rather than base, collector, and emitter. Applying a voltage to the gate controls the current between the source and drain. FETs come in two types: N-channel (similar to NPN) and P-channel (similar to PNP).

Technically, FETs come in two sub types: MOSFET and JFET. For the purposes of your basic electronics education, the differences between these two types don't really matter, but knowing this kind of secret electronics language helps you sound smarter when you talk to your electronics geek friends.

Static discharge can damage FET transistors. At the very least, always store your FETs in anti-static foam. When buying FETs, keep them in their anti-static bag or tube and leave them there until you're ready to use 'em.

Packing Parts Together on Integrated Circuits

All the components that we mention in the earlier sections of this chapter come just one to a package. Electronics mavens call them *discrete* components, meaning separate. (Don't confuse the word with *discreet* which means minding your own business.)

Enter the integrated circuit, the true marvel of the 20th century. Also called a chip or IC, these amazing creations are miniature circuit boards produced on a single piece of semiconductor. A typical integrated circuit contains hundreds of transistors, resistors, diodes, and capacitors. Because of this circuit efficiency, you can build really complex circuits with just a couple of parts. ICs are the building blocks of larger circuits. You string them together to form just about any electronic device you can think up.

The way that all the components are wired inside an IC determines what the IC does. You can either solder the IC directly into the circuit board or mount it in a socket.

Integrated circuits most often come enclosed in dual in-line pin (DIP) packages, such as the ones in Figure 4-13. This illustration shows several sizes of DIP ICs, from 8-pin to 40-pin. The most common sizes are 8-, 14-, and 16-pin.

Linear, digital, or combination plate?

Over the years, chip makers have come out with thousands upon thousands of different ICs. Each one does something special. Many of the integrated circuits you encounter are standardized, and you can read various books to discover more about them. A lot of chip makers offer these standardized ICs, and manufacturers and electronics hobbyists the world over buy and use them in various projects. Other ICs, called special-purpose ICs, are designed to accomplish some unique task. More often than not, only a single company sells a particular special-purpose chip.

Whether standardized or special-purpose, you can separate ICs into two main categories: linear and digital. These terms relate to the kinds of electrical signals that work within the circuit:

- ✔ **Linear ICs:** These ICs are designed to work with any circuit that uses varying voltages and currents (an analog circuit). An example of an analog circuit is a guitar amplifier.

- ✔ **Digital ICs:** These ICs are designed to work with a circuit that uses just two voltages (a digital circuit). As we note in Chapter 1, these two voltages indicate binary digital data (on/off, high/low, 0/1, that sort of thing). Common voltages that represent digital data are 0 and (often) 5 volts. Refer to Chapter 4 for more detail about digital circuits and binary data.

The majority of standardized ICs fall into either the linear or digital category. Most mail order outfits that sell ICs separate them into linear and digital lists. Some ICs are made to work with both analog (linear) and digital signals, and some can convert between digital and analog signals or work with a host of other combinations. There's no sense in trying to corral all of the variations in this book, except to note that you can't neatly classify all chips as either linear or digital.

Figure 4-13: Among the most popular form-factors for integrated circuits is the dual in-line pin (DIP) package.

IC part numbers

ICs — like transistors, — have a unique number code to identify them. This code, such as 7400 or 4017, indicates the type of device. You can look up specifications and parameters about an IC in a reference book or online. The code is printed on the back of the IC.

Many ICs also contain other information, including manufacturer catalog number and maybe even a code that represents when the chip was made. Don't confuse the date code or catalog number with the code that the catalog uses to identify the device. Manufacturers don't have any standards for how they stamp the date code on their integrated circuits, so you may have to do some detective work to pick out the actual part number of the IC.

Understanding IC pinouts

By their nature, integrated circuits require multiple connections to a circuit. These connections are called *pins*. One pin may be for power, another for ground, another for input, yet another for output, and so forth. The function of each pin is referred to as *pinout*. The pinout isn't printed on the top of the integrated circuit. In order to use the IC in a project you have to look up the pinout in the data sheet for the integrated circuit. You can find these data sheets for most common (and many uncommon) ICs on the Internet. Use a Google or Yahoo! search to help you locate them.

In order to identify what each pin is for, by convention, the pins on an IC are numbered counterclockwise, starting with the upper-left pin closest to the clocking mark. The *clocking mark* is usually a notch, but it can also be a little dimple, or white or colored stripe. The pins are numbered looking down from the top of the IC, starting from 1. So, for example, the pins of a 14-pin IC are numbered 1 through 7 down the left side and 8 through 14 up the right side, as you can see in Figure 4-14.

Schematic diagrams show the connections to integrated circuits in one of two ways:

- Some schematic diagrams show an outline of the IC with numbers beside each pin. The numbers correspond to the clocked pinout of the device. (Remember, start with 1 in the upper left and go counterclockwise.) You can easily wire up an IC with these kinds of diagrams because you don't need to look up the device in a book or data sheet. Just make sure that you follow the schematic and that you count the pins properly.

- If the schematic lacks pin numbers, you need to find a copy of the pinout diagram. For standard ICs, you can find these diagrams in reference books and online; for non-standard ICs, you have to visit the manufacturer's Web site to get the data sheet.

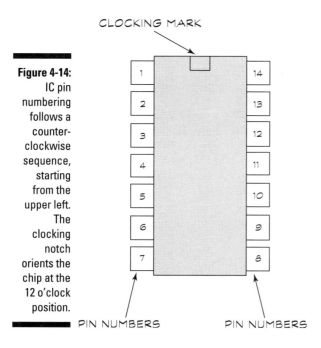

CLOCKING MARK

Figure 4-14: IC pin numbering follows a counter-clockwise sequence, starting from the upper left. The clocking notch orients the chip at the 12 o'clock position.

PIN NUMBERS   PIN NUMBERS

You can always make a reference copy of the pinout, even if the schematic includes the pin numbers. With this copy, you can double-check your work (and the schematic) to help ensure accuracy. The schematic may have numbered the pins incorrectly, and you can save yourself a lot of trouble and frustration by checking the schematic against the pinout diagram.

Exploring ICs on your own

There's more to integrated circuits than we can possibly cover in this book. If you're interested in learning more, see the Appendix. You'll find interesting Web sites that provide useful how-to info for using various popular ICs in working projects.

Chapter 5

Filling Out Your Parts Bin

. . . .

In This Chapter

▶ Picking the perfect type of wire

▶ Powering up with batteries and solar cells

▶ Flipping switches

▶ Controlling output with logic gates

▶ Tuning signals with inductors and crystals

▶ Making sense of things with sensors

▶ Exploring how DC motors work

▶ Making some noise with speakers and buzzers

. . . .

*A*lthough the resistors, capacitors, diodes, and transistors that we discuss in Chapter 4 are pretty darn important (you'd have trouble finding
a circuit in the world that you can build without them), you need to know
about some other parts that you make use of in your electronics career.

Some of these other parts, such as wires, connectors, and batteries are pretty
essential. After all, you'd be hard put to build an electrical circuit without
wires to connect things together or a source of power to make things run.
You use some other parts that we discuss in this chapter only now and then
for certain circuits; for example, you don't want to use one of those annoying
buzzers on every circuit that you build, but when you want to make noise,
they come in handy.

In this chapter, we discuss a mixed bag of parts, some of which you need to
stock up on right away, and others you can leave until you need them.

Making the Connection

Making a circuit requires that you connect components to allow electric current to flow between them. The following sections describe wires, cables, and
connectors that allow you to do just that.

Wire

Wire that you use in electronics projects is just a long strand of metal, usually made of copper. The wire has only one job — to allow electrons to travel through it. However, you can find a few variations in the types of wire available to you. In the following sections, we talk about which type of wire you use for different situations.

Stranded or solid wire?

Cut open the cord of any old household lamp (make sure that you're ready to junk it and unplug it first!!), and you see two or three small bundles of very fine wires, each wrapped in insulation. This is called *stranded* wire. If, on the other hand, you have only one wire on its own, instead of a bundle of wires, you call it *solid* wire. You can see examples of stranded and solid wires in Figure 5-1.

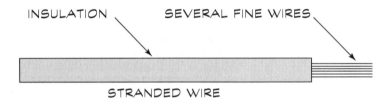

INSULATION   SEVERAL FINE WIRES

STRANDED WIRE

Figure 5-1:
Do you think
wire is just
wire? Think
again. Here
are two
types.

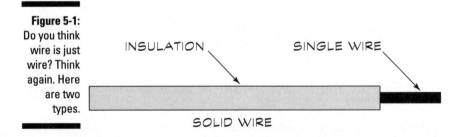

INSULATION   SINGLE WIRE

SOLID WIRE

When do you use each type of wire? It's not as complicated as you may think: You use stranded wire in projects where the wire will be moved around. For example, you use stranded wires for multimeter leads because you move and flex the leads frequently. If you use a solid wire, it snaps in two after you flex it several times.

Use solid wire to connect components on breadboards (check out Chapter 11 for more on breadboards) and other places where you don't plan to move the wire around. You can easily insert the solid wire into holes in the board, and that wire stays in the shape to which you mold it. If you try to use a stranded wire in a breadboard, you have to twist the strands to get all of them in the hole, and you may break a strand which could short out the circuit.

Size matters

You refer to the size of wire as wire gauge. The *wire gauge* is simply short-hand for the diameter of the wire. What's confusing is that the relationship between wire gauge and wire diameter is essentially backwards. A smaller wire gauge means a larger wire diameter.

Manufacturers saddled us with this backwards-naming scheme because of the manufacturing process they use for wires. To make a wire, the metal (usually copper) is pulled through a hole in a steel plate. To make a small diameter wire, the wire is pulled through a series of holes, each hole smaller than the previous one. The wire gauge refers to the number of different size holes the wire was pulled through to make the desired diameter. So the higher the number, the more times someone had to pull the wire and the smaller that wire got in the process. You can see common wire gauges in Table 5-1.

Table 5-1	Wires Commonly Used in Electronics Projects
Wire Gauge	*Wire Diameter (inches)*
16	0.051
18	0.040
20	0.032
22	0.025
30	0.01

Here are a few gauge guidelines for you:

✔ You can use 20- or 22-gauge wire for most electronics projects.

✔ You may find 16- or 18-gauge wire useful for heavy-duty applications, such as connecting motors to a power supply.

✔ You use 30-gauge wire for *wire wrapping,* a method for connecting components on circuit boards, as we discuss in Chapter 12.

The wire gauges listed here work for the types of projects that we cover in Chapters 14 and 15. You use smaller gauge (and therefore larger diameter) wires for many other types of heavy-duty applications. For example, if you have an electric range in your kitchen, it typically uses 6-gauge wire (which has a diameter of 0.162 inches) to supply electric current to the range.

Note that you sometimes see gauge abbreviated in weird and wonderful ways. For example, you may see 6 gauge abbreviated as 6 ga., #6, or 6 AWG (AWG stands for American Wire Gauge).

If you start working on projects in which you use higher voltage or currents than the ones we describe in this book, consult the instructions for your project or an authoritative reference to determine the correct wire gauge. For example, the *National Electrical Code* lists the appropriate wire gauges for each type of wiring that you use in a house. Make sure that you also have the right skills and sufficient knowledge of safety procedures to work on such a project.

The colorful world of wires

Whoever said that electronics isn't a colorful subject didn't know a thing about wire. The insulation around wire comes in different colors to help you identify what you should attach the wire to.

Look at the connector for a 9-volt battery, for example. You see one red and one black wire. The red wire attaches to the positive terminal of the battery, and the black wire attaches to the negative terminal.

When wiring up a circuit (for example, when you work with a breadboard), you use certain color wires for each type of connection. This distinction allows you (if you don't have a photographic memory) or someone else to look at the circuit and identify each type of connection.

Here are the different colors and their suggested uses:

- Use red wire for all connections to +V (positive voltage).

- Use black wire for all connections to –V (negative voltage) or ground. If you're connecting the wire to ground, you can also use green wire.

- Use yellow or orange wire for input signals, such as the signal from a microphone. If you have more than one input signal, use a separate color for each.

Collecting wires into cables or cords

Cables are actually groups of two or more wires protected by an outer layer of insulation, such as the power cord mentioned earlier in the section "Stranded or solid?". Cables differ from stranded wires because the wires used in a cable are separated by layers of insulation. Cables are less fragile than individual wires, so you can string them between pieces of equipment — for example, you can use cables to connect a TV to a DVD player. You can see a cable with plug-in connectors in Figure 5-2.

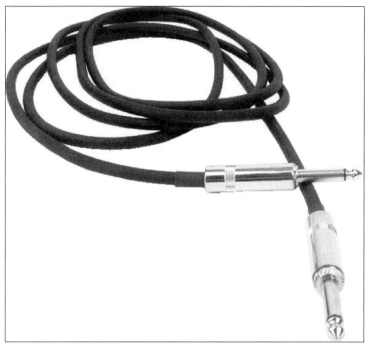

Figure 5-2:
A cable with
plug-in
connectors.

Making connections with connectors

If you take a look at a typical cable, say the one that goes from your computer to your printer, you see that it has metal or plastic doodads on each end. There are also metal or plastic receptacles on your computer and printer that these cable ends fit into. You call all of these metal or plastic parts *connectors.* You insert one connector on the cable, called a *plug,* into a matching connector (in this case, a socket or *jack*) on the printer. The various pins and holes connect the appropriate wire in the cable to the corresponding wire in the device.

Here are the connectors that you run into most often in your electronics adventures:

✔ A *terminal and terminal block* work together as the simplest type of connector. A terminal block contains sets of screws in pairs. You attach the block to the case or chassis of your project. You then solder (or crimp) a wire to a terminal for each wire that you want to connect. Next, you connect each terminal to a screw on the block. To connect wires to each other, simply pick a pair of screws and connect the terminal on each wire to one of those screws. For many of the early projects that you tackle, you don't need anything more complex than this type of connector.

- ✔ A useful *terminal-block variation* mounts the terminal block on a printed circuit board. This type of terminal block allows you to simply insert the bare end of a wire into a contact in the block, instead of soldering a terminal to the wire.

- ✔ *Plugs* and *jacks* that carry audio signals between equipment, as from a guitar to an amplifier, have cables such as the one you see in Figure 5-2. There are plugs on both ends of the cable and a jack is mounted on both the guitar and amplifier. These cables contain either one or two signal wires; a metal shield surrounds the wires. This metal shield minimizes the introduction of current into the wires, limiting noise that can disrupt a signal.

- ✔ You typically use *pin headers* to bring signals to and from circuit boards (we talk more about circuit boards in Chapter 12). You mount the socket half of this kind of connector on the circuit board, and you attach the plug half to a ribbon cable. The rectangular shape of the connector allows for easy routing of signals from each wire in the cable to the correct part of the circuit board. You refer to these connectors by the number of pins — for example, you may talk about a 40-pin header. After you start building robots or other more complex projects that involve more than one circuit board, you can find a use for this type of connector.

Electronics uses many connectors that you don't need to know about until you get into more complex projects. When you build that spaceship, or any other more complicated gadget, you can look up details on the connectors that you need on many electronic suppliers' Web sites or in catalogs.

Powering Up

All the wires and connectors in the world won't do you much good if you don't have a power source. After you build a project, you need voltage and current to get the thing going. You can get power from your wall outlet (we talk about plugging into your wall outlet in Chapter 3), batteries, or solar cells.

For electronics projects, batteries and solar cells make great power sources because they're lightweight and portable. The following sections discuss how to choose batteries and solar cells for your projects.

Turning the juice on with batteries

A battery uses a process called an electrochemical reaction to produce a positive voltage at one terminal and a negative voltage at the other terminal. This process involves placing two different types of metal in a certain type of chemical. (Because you didn't buy *Chemistry For Dummies* — by John T. Moore from Wiley Publishing, by the way — we don't get into the guts of a battery here, just the basics.)

You can categorize batteries by size, voltage, and the type of chemicals that they contain, such as zinc-carbon or nickel-cadmium. Figure 5-3 shows a few typical battery sizes.

Starting with your everyday-type batteries

Start with the standard, non-rechargeable type of batteries that you can buy in any supermarket. The AAA-, AA-, C-, and D-size batteries all produce about 1.5 volts, compared to the transistor battery (that little rectangular battery that looks sort of like a Lego® block found in lots of small electronic gadgets) that produces about 9 volts and the *lantern battery* (that big boxy thing that fits in flashlights the size of a boom box) that produces about 6 volts.

You can combine any number of 1.5-volt batteries to get the voltage that you need for your project. For example, when you connect the positive terminal of one battery to the negative terminal of another battery (you call this set-up connecting the batteries in series), as in Figure 5-4, you get twice the voltage.

Figure 5-3:
Getting alphabetical with batteries in small (AA), medium (C) and large sizes (D).

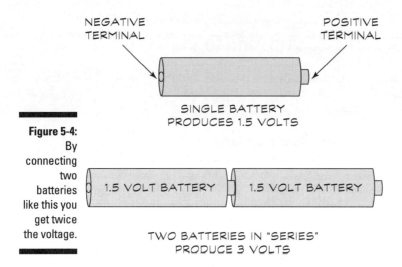

NEGATIVE
TERMINAL

POSITIVE
TERMINAL

SINGLE BATTERY
PRODUCES 1.5 VOLTS

1.5 VOLT BATTERY   1.5 VOLT BATTERY

TWO BATTERIES IN "SERIES"
PRODUCE 3 VOLTS

Figure 5-4:
By connecting two batteries like this you get twice the voltage.

You put batteries together in battery holders. When you place four 1.5-volt batteries in a battery holder, for example, those batteries combine to produce 6 volts; when you place six 1.5-volt batteries in a battery holder, they combine to produce 9 volts; and so on. Figure 5-5 shows a battery holder containing four AA batteries.

Figure 5-5:
Four 1.5-volt batteries tucked into a battery holder produce about 6 volts.

Taking AA batteries to the max

The amp-hour or milliamp-hour rating for a battery gives you a measure of how much current a battery can conduct for a given length of time. For example, a 9-volt transistor battery usually has about a 500 milliamp-hour rating (this measurement varies with the battery type; see the section "Sorting batteries by what's inside" for more about battery types). Such a battery can power a circuit using 25 milliamps for approximately 20 hours before its voltage begins to drop. An AA battery may have a 1500 milliamp-hour rating. Therefore, a battery pack containing AA batteries can power a circuit using 25 milliamps for approximately 60 hours.

For many projects that require a 9-volt supply, you actually do better using a battery holder with six smaller voltage AA batteries than a single 9-volt battery. Why? The AA batteries last longer than the single 9-volt battery. The amount of electric current that a battery can generate before it depletes the chemicals it contains varies. A battery holder with six AA batteries cumulatively contains more chemicals than a single 9-volt battery, and so the battery holder lasts longer. (This example assumes that both batteries use the same chemicals, which we discuss later, in the section "Sorting batteries by what's inside"). When you use a battery it begins to wear out and the voltage drops; for example we checked a 9-volt battery that we'd used for a few days and found that it was only producing 7-volts.

Batteries that just keep on going

If you have a project that uses a lot of current, or you plan to run the gadget all the time, it can eat through non-rechargeable AA batteries like you go through popcorn at a movie theater. In that case, you can use

- **C or D size batteries:** These batteries are bigger than AA batteries; remember, the bigger the battery, the longer it lasts.

- **Rechargeable batteries:** Some batteries allow you to revitalize the chemicals that they use, bringing them back to something like their original charge. See the next section for more about rechargeable batteries.

Though some (fool) hardy souls do recharge non-rechargeable batteries, it's not a good idea. The batteries can rupture and leak acid, or worse (think exploding batteries — not a pretty picture).

Sorting batteries by what's inside

Batteries are classified by the chemicals they contain. Note that any of the various size batteries that we discuss in the previous sections can contain these chemicals, and the chemicals in a battery relate to whether that battery is rechargeable or non-rechargeable.

If you buy rechargeable batteries make sure that the battery charger you use is designed for that type of rechargeable battery.

Some of the readily available battery types use the following chemicals:

- **Zinc-carbon:** This type of battery falls at the low quality end of non-rechargeable batteries. Although they don't cost very much, you have to replace them frequently.

- **Alkaline:** We suggest that you start with this type of battery for your projects. These batteries last about three times as long as zinc-carbon batteries. When you find yourself doing so many projects that you need to replace these batteries frequently, step up to rechargeable batteries.

- **Nickel-cadmium (Ni-Cad or NiCad):** This is the most popular type of rechargeable battery. Though many manufacturers have eliminated the problem today, the big flaw with some nickel-cadmium batteries is something called the memory effect. With the memory effect, you need to fully discharge the battery before recharging it to insure that it recharges to full capacity. If you don't discharge it, it doesn't charge fully. Nickel-cadmium batteries generate about 1.2 volts.

- **Nickel-metal hydride (Ni-MH):** This type of rechargeable battery generates about 1.2 volts. This battery doesn't suffer from the memory effect seen in nickel-cadmium batteries. If you decide to use rechargeable batteries, we suggest that you start with these. Buying a recharger and a supply of these batteries saves you a considerable amount of money over time.

- **Lithium:** If you're working on a project that requires a lightweight battery, consider lithium. This type of battery generates higher voltage than other types, at about 3 volts. Lithium also has a higher capacity than alkaline batteries. They cost more, and you can't recharge most batteries of this type. But for a project where you need to watch your weight (no, we don't mean dieting), such as when moving a small robot around the house, you may find them very useful.

Don't worry about whether to use lithium-polymer or lithium-ion batteries. Some battery experts speculate that the manufacturing process for lithium-polymer batteries may evolve to produce a better battery in the future. However, at this point in time, you can't really make a strong case for them having any advantages over the lithium-ion type. So just go with the battery that you can find most easily, or the one that costs less.

Turning on power with solar cells

In Chapter 4, we discuss light-emitting diodes that generate light when you apply an electric current to them. Conversely, if you shine light on diodes,

they generate an electric current. A *solar cell* is just a large diode that generates current when exposed to a light source, such as the sun.

To power a project, you can buy panels of solar cells. You have to weigh the
voltage and current requirements for your project against the size of the
solar panel. For example, a panel measuring about 5 x 5 inches may be able
to generate 100 milliamps at 5 volts in bright sunlight. If you need 10 amps,
you can get it, but you may find the size of the panel problematic on a smaller
or more portable project.

Look at the following criteria before choosing a solar panel for your project:

✔ Do you plan to have the solar panel in sunlight when you want the
gadget to be on?

If not, look for another power source. Or, if you want to get fancy, design
the gadget so that the solar cell provides a charge to the batteries to
power the gadget even when it's dark.

✔ Is a solar panel that provides enough power small enough to fit on your
gadget?

If not, redesign the gadget to take less power or look for another power
source.

Turning Electricity On and Off

You've scrounged around your growing electronics bin and come up with
wires to connect a circuit together and batteries to power the circuit. So how
do you turn the power on and off? You use switches and relays, which we
cover in the following sections.

Turning current on and off with switches

When you move the switch to shut off your flashlight, you disconnect the
wires that run from the battery to the light bulb. All switches do the same
thing: Connect wires to allow electric current to flow or disconnect wires to
stop electric current from flowing.

When you turn off your flashlight, you put the switch in what is called the
open position. With the switch in the *open position,* you have a disconnected
wire, and no current can flow. When you turn on the flashlight, you put the
switch in the closed position. With the switch in the *closed position,* you've
connected the wire (and completed the circuit), and current can flow.

Starting with simple switches

Your flashlight usually comes with something called a slide switch. With a *slide switch,* you slide the switch forward or backward to turn the light on or off. You can see some other types of switches (toggle, rocker, and leaf switches) in Figure 5-6.

Figure 5-6: From top to bottom: two toggle switches, a rocker switch, and a leaf switch.

Toggle, rocker, and slide switches all do the same job, so grab whatever switch you have handy that you can easily use on the project that you're building. For example, a slide switch works well on a round, handheld flashlight because of the position of your thumb, but a toggle switch may work best to flip on a gadget sitting on your workbench.

Want to see a leaf switch in action? In chapter 15 we describe how you can use a leaf switch like a car bumper that tells a robot when it's bumped into something. Push-button switches come in three versions:

- ✔ **Normally closed (NC):** This push-button switch disconnects the wire only when you push the button.

- ✔ **Normally open (NO):** This push-button switch connects the wire only when you push the button.

- ✔ **Push on/Push off buttons:** This switch connects the wire with one push and disconnects the wire with the next.

You typically find push-button switches in electronics to start or stop a circuit. For example, you press a normally open push-button switch to ring a doorbell.

What's inside a switch?

You call the basic switches that we talk about in the previous section single-pole single-throw, or SPST types. Don't worry about all the different names: In essence, these switch types have one wire coming into the switch and one wire leaving it.

Just to keep your electronics life interesting, you may come across other types of switches that are wired a bit differently, called double pole. Where *single pole switches* have one input wire, *double pole switches* have two input wires. With single throw switches you can connect or disconnect each input wire to one output wire, while double throw switches allow you to choose which of two output wires you connect each input wire to.

There are a few single- and double-pole variations, including

- **Single-pole double-throw (SPDT):** In this switch, one wire comes into the switch and two wires leave the switch. When you want to choose what device a circuit turns on (for example, a green light to let people know that they can enter a room or a red light to tell them to stay out), use an SPDT switch.

- **Double-pole single-throw (DPST):** This switch has two wires coming into it and two wires leaving. You can use a DPST switch to control two separate circuits. For example, you can have one circuit with components running on 5 volts and another circuit with components running on 12 volts. With one switch, you can turn both circuits on or off.

- **Double-pole double-throw (DPDT):** This switch has two wires coming into it and four wires leaving. A DPDT switch has three positions. In the first position, the first pair of output wires connect. In the second position, all four output wires disconnect (some DPDT switches do not have this position). In the third position, the second pair of output wires connect. You can use this type of switch to reverse the polarity of DC voltage going into a motor so that the motor turns in the opposite direction. (One position makes the motor turn clockwise, one position turns off power to the motor, and one position turns the motor counterclockwise.)

Let a relay flip the switch

You've made a gadget to let you know when your no-good brother-in-law, Herman, is raiding the refrigerator. But there's one problem: The gadget runs on a 5-volt battery pack, and you want the gadget to turn on enough sound and light to scare the guy into the next county. No problem, just use a relay.

How relays work

A *relay* is simply an electrically powered switch. When your gadget sends 5 volts to the relay, an electromagnet turns on and then closes a switch inside the relay. If you wire that switch to 117 volts, you can power enough lights and sirens to send Herman scurrying.

Exploring electromagnets

So how does the electromagnet part of a relay setup work? An *electromagnet* can be something as simple as coiled wire around an iron bar or even a nail. When you run some current through the wire, the bar becomes magnetized. When you shut off the current, the bar loses that magnetic quality.

Two magnets attract or repel each other, depending on which ends (or poles) of the magnets you put together. Part of the switch contained in a relay consists of a lever attached to a magnet, as you can see in Figure 5-7. When voltage runs through the wire coil, the electromagnet pulls the lever toward it, and the switch closes, connecting the 115 volts to the lights and sirens (goodbye, Herman!). When you shut off current to the wire coil the electromagnet shuts off and a spring pulls the lever away, opening the switch.

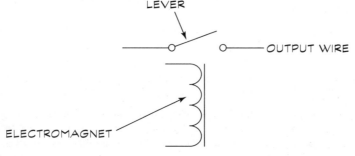

LEVER

OUTPUT WIRE

Figure 5-7:
Here's how
the parts of
a relay fit
together.

ELECTROMAGNET

You can find relays that use 5, 12, or 24 VDC to power an electromagnet with a SPST, SPDT, or DPDT switch (see "What's inside a switch," earlier, for more info about switch types).

Often, instead of saying that a switch in the relay opens or closes, people talk about contacts opening or closing. Also, people sometimes call a lever in a relay an armature. But a relay by any other name, would work the same...

Making Decisions with Logic Gates

If you've ever played computer (or even old-fashioned) chess, you know that the game takes a little simple logic. When a knight threatens your rook, there are only so many moves that you can make. You have to go through the possible moves in your mind to decide which one can save your rook.

So if your opponent is a computer, how the heck did the machine learn to be logical?

Skipping the complicated programming and electronic circuits involved in a computer chess game, just what enables programming and circuits to implement logic? That trick involves something called logic gates. *Logic gates* are integrated circuits that take input values and determine what output value to use based on a set of rules. Usually, logic gates have two inputs; one type of logic gate, called an *inverter,* has only one input. You can even get logic gates with more than two inputs, which you may need for some projects.

See Tables 5-2 through 5-6 in the section "Common logic gates," later in this chapter, for information about how output varies depending on input.

Using logic in electronics

Although you may not build the next generation of computer chess games, you can use logic gates for simpler things. For example, the microprocessor in the calculator you use to balance your checkbook uses logic gates to add, subtract, multiply, and divide. On the level of electronics projects that you're likely to tackle at first, you can build a simple circuit with some logic gates to count how many times the door to your house opens, monitoring your family's comings and goings.

When people talk about logic gate input and output values, they say that an input is high (1) or low (0 – zero). In a typical circuit, *high* means the circuit has a voltage of approximately 5 volts because that is the voltage typically used to turn on a transistor. *Low* means it has the voltage of approximately zero volts.

Common logic gates

You may encounter any of these five common logic gates: AND, OR, Inverter (or NOT), NAND (basically an AND gate followed by an inverter), and NOR (basically an OR gate followed by an inverter). Tables 5-2 through 5-6 show you the output for each of these gates with various combinations of inputs.

The name of each logic gate comes from how the inputs determine the output. For example the output of the AND gate is high only when both inputs (one input AND the other input) are high, but the output of the OR gate is high when either one OR the other, OR both of the inputs is high. We show you the symbols used for each type of gate in Chapter 6.

Table 5-2	AND Gate Input/Output	
Input A	*Input B*	*Output*
Low	Low	Low
Low	High	Low
High	Low	Low
High	High	High

Table 5-3	OR Gate Input/Output	
Input A	*Input B*	*Output*
Low	Low	Low
Low	High	High
High	Low	High
High	High	High

Table 5-4	Inverter Input/Output
Input	*Output*
Low	High
High	Low

Table 5-5	NAND Gate Input/Output	
Input A	*Input B*	*Output*
Low	Low	High
Low	High	High
High	Low	High
High	High	Low

Table 5-6	NOR Gate Input/Output	
Input A	*Input B*	*Output*
Low	Low	High
Low	High	Low
High	Low	Low
High	High	Low

You usually find multiple logic gates sold in integrated circuits, such as an IC containing four two-input AND gates (called a quad 2-input AND gate). Look on the Web site of your IC's manufacturer for a data sheet that tells you which pins are inputs, outputs, V+ (voltage), and ground. This information allows you to make choices, such as whether to use one or all four of the logic gates in the integrated circuit.

Make sure that the part you buy has the number of inputs that you need for your project. Remember that you can buy logic gates with more than two inputs. For example, you can find a 3-input AND gate from most electronics suppliers.

Some components already have gates wired together in a circuit to perform functions, such as counting or decoding. When you work with circuits that require such functions, look at the manufacturer's datasheet to determine how the part works and to understand the nature of each pin.

Controlling Frequency with Inductors and Crystals

Inductors and crystals both have a relationship with frequency. Inductors are used to weed out all but a desired frequency (this is one of the pieces of the process when a radio tunes into only one station; more about this shortly). Crystals, on the other hand, are often used to generate specific frequencies in a circuit.

Storing energy in inductors

Anybody who has driven across country and experienced the annoyance of tuning into and then losing a radio station every ten minutes knows that radio stations come and go. But exactly what goes on inside your average radio to bring you those tunes (however fleeting)?

An inductor by any other name

You sometimes hear inductors called by other names, including choke and coil. You can use these names interchangeably, but they don't always have the same meaning to circuit designers. For example, designers most often use a coil to resonate at a certain frequency, and they use a choke to reject a group or range of frequencies. We use the word inductor throughout this book to encompass all the little sub-varieties.

Every radio station broadcasts electric waves at a particular frequency. When you change radio stations, you're actually tuning your radio into a new frequency by changing variable components in the circuit so that the radio allows only signals at that frequency through.

So what exactly weeds out all signals but your favorite Top 20 station? Circuits use inductors (you may hear inductors also called "coils" or "chokes") along with capacitors (which we discuss in Chapter 4) to filter out all but one frequency.

You also find inductors in many other types of circuits. For example, more elaborate power supplies use inductors as a means of reducing the 60-Hz "noise" that often occurs on an incoming power line.

Inductance is the ability of a wire coil to store energy in the magnetic field that surrounds it when current is flowing through the wire. You state the value of an inductor in henrys (H or h), or more commonly, millihenrys (thousandths of a henry) and microhenrys (millionth of a henry). The value of an inductor measures its ability to reduce the voltage of an AC signal. The value of inductors is typically marked using the same color-coding technique used for resistors, which you can read about in Chapter 4. You can often find the value of larger inductors printed directly on the components. Smaller value inductors look a lot like low-wattage resistors, and these inductors and resistors even have similar color-coding marks. Larger value inductors come in a variety of sizes and shapes depending upon their application.

Inductors can be either fixed or variable. With both types, a slender wire winds around an insulating core. The number of turns of the wire, the core material, and the wire's diameter determine the value of the inductor. Fixed inductors have a constant value, while variable inductors sport a knob you can turn to adjust the value.

The core of an inductor can be made of air, iron ferrite, or any number of other materials (though inductors use air and ferrite most often).

Making frequencies crystal clear

When you cut a quartz crystal to a certain thickness, that crystal resonates at a particular frequency. You use crystals in circuits called oscillators to generate electric signals at a certain frequency. Microcontrollers, such as the BASIC Stamp controller (we talk about this device in Chapter 13), use oscillators. Many other types of electronic gadgets make use of oscillators, too.

You state the frequency for a crystal in MHz (Megahertz, a measurement that you can read more about in Chapter 1). Crystals have two leads to connect them to a circuit, and you can buy them in a variety of shapes.

This may seem obvious, but when you buy a crystal, be sure to get one with the operating frequency appropriate for your project.

Making Sense of Things

You use some components to trigger a circuit to do something (such as turning on a light) when those components sense that some event has happened (such as a change in temperature). Handily enough, these are called sensors. The following sections cover a few sensors that you may come across in your electronics projects.

Can you see the light?

There's one variation among several of the standard components that you can read about in Chapter 4: Light dependency. Manufacturers make certain versions of resistors, diodes, and transistors sensitive to light. The output of these components varies as the amount of light shining on them changes. You can use light-dependent components as sensors in equipment such as burglar alarms, safety devices that work with your descending electric garage-door to stop it when a cat runs underneath, and automatic dusk-to-dawn lighting. You can also use them for communication. If you use a remote control for your TV (and who doesn't?), your TV contains a light-sensitive transistor or diode to receive signals from that remote control.

Here is the lowdown on how resistors and transistors fit into the light sensor equation:

- ✔ You can refer to light-dependent resistors as photocells or photo resistors; their resistance changes based on the amount of light falling on them. The typical photocell is most sensitive to visible light, especially in the greenish-yellow spectrum.

> ✔ You can refer to light-dependent transistors and diodes as phototransistors and photodiodes, respectively. From the outside, phototransistors and photodiodes look identical to each other, so make sure to keep them in separate cubbies in your parts bins. Both are sensitive mostly to infrared light (essentially this is heat) which you cannot see. When you use a remote control for your TV you are using a photodiode in the remote to send infrared signals to a phototransistor in your TV.

We give you only the basics about components that work by light in this chapter. Take a look at Chapter 14 for some nifty hands-on projects using several types of light-dependent components.

Sensing the action with motion detectors

When you walk up to someone's front door and the light turns on, you've found a motion detector at work. Many homes, schools, and stores use motion detectors to turn on lights or detect burglars.

Most motion detectors use a technique called passive infrared (PIR) and use or control 117-volt circuits. Typically, these models mount on a wall or on top of a floodlight, and they take up a lot of room.

For a project using a battery pack, you probably need a compact motion detector that works with about 5 volts. You can find this kind of motion detector through online security system vendors.

Getting inside a motion detector's head

The insides of a PIR motion detector actually are fairly simple. PIR motion detectors contain two crystals, a lens, and a small electronic circuit. When infrared light (basically, heat which you or any other warm thing generates) hits the crystal, it generates an electric charge. A person gives off heat, as do other living things, so you set off a motion detector when you come near it.

A typical motion detector has three wires: ground, positive supply voltage, and the output for the sensor. If you supply +5 volts to the PIR, the voltage on the output wire reads about 0 (zero) volts when the PIR detects no motion. When it does detect motion, the voltage on the output wire reads about 5 volts.

Don't buy a PIR sensor rather than a motion detector. A sensor doesn't have the lens that comes with a motion detector. It's that lens that helps the sensor to detect the motion of something rather than just the presence of something.

Varying the motion: Other types of motion detectors

In case you have a special interest in surveillance and want to know more about motion detectors, we take a moment here to give you the lowdown on a couple of other types:

- **Active infrared motion sensor:** This sensor uses an LED that emits infrared light and a detector, such as a phototransistor, that generates current when infrared light strikes it. When someone passes between the LED and the sensor, the phototransistor stops generating current. This sensor is really just a version of the old-fashioned electric eye that readers over 40 might remember from old James Bond movies, and you can use it effectively only in areas with regular traffic, such as hallways.

- **Ultrasonic motion detector:** This detector generates ultrasonic waves that reflect off any objects in the room. If nothing in the room moves, the ultrasonic waves bounce back with no change. If someone or something moves in the room, the ultrasonic waves distort, and that distortion triggers an alarm. You don't really have a compelling reason to use these devices rather than the PIR detectors unless you have a special fondness for ultrasonic gadgets.

You're getting warmer: Temperature sensors

Remember when you were a kid lying in bed on a cold winter evening? Suddenly, you heard a sound! But you quickly realized that it wasn't the boogeyman coming to get you — just the furnace turning on in your cold house. The thermostat in your wall activated the furnace because it sensed that the temperature had dipped below the preset temperature.

A thermostat uses a coiled metal strip (called a *bimetallic strip*) that shrinks as the temperature cools. When the coil shrinks to the point you set on your thermostat, it trips a switch and turns the furnace on. This is a simple and common type of temperature sensor that you use in certain types of gadgets, such as the thermostat. Other types of temperature sensors, including thermocouples, semiconductor temperature sensors, infrared temperature sensors, and thermistors measure changes in temperature electrically, rather than mechanically, as with the bimetallic strip.

To make your life a little easier, we suggest that you just stick to using thermistors for projects where you want to measure temperature because they are generally easier to use than thermocouples and infrared temperature sensors. A *thermistor* is a resistor whose resistance value changes with changes in temperature.

Other ways to sense temperature

In the section "You're getting warmer: Temperature sensors," we mention several other types of temperature sensors. Here, we briefly summarize their characteristics for the curious among you:

✔ **Semiconductor temperature sensors:** After thermistors, these are probably the easiest to use. The most common type of this sensor contains two transistors. This sensor's output voltage depends on the temperature.

✔ **Thermocouples:** These sensors generate a voltage that changes with temperature. Thermocouples contain two wires (for example, a copper wire and a wire made of a nickel/copper alloy) welded or soldered together at one point. The metals it uses determine how the voltage changes with temperature. You can use thermocouples to measure high temperatures, such as several hundred degrees or even over a thousand degrees.

✔ **Infrared temperature sensors:** These sensors measure the infrared light given off by an object. You can use these sensors when the sensor must stay at a distance from the object you plan to measure; for example, you use this sensor if a corrosive gas surrounds the object. Industrial plants and scientific labs typically use thermocouples and infrared temperature sensors.

There are two types of thermistor:

✔ **Negative temperature coefficient (NTC) thermistors:** The resistance of this type of thermistor decreases with a rise in temperature.

✔ **Positive temperature coefficient (PTC) thermistors:** The resistance of this type of thermistor increases with a rise in temperature.

You should find an NTC or PTC marking on your thermistor; if you can't find this marking, you can verify which type of thermistor you're dealing with when you calibrate it by identifying whether the value increases or decreases with a rise in temperature.

Suppliers' catalogues typically list the resistance of thermistors at 25 degrees Celsius (77 degrees Fahrenheit). Measure the resistance of the thermistor yourself with a multimeter (see Chapter 9 for more about using multimeters) at a few temperatures; these measurements give you the resistance at each temperature so that you can calibrate the thermistor. If you plan to use the thermistor to trigger an action at a particular temperature, make sure to measure the resistance of the thermistor at that temperature. Thermistors have two leads and no polarity, so you don't need to worry about which lead you have wired to +V.

Good Vibrations with DC Motors

Have you ever wondered what causes a pager to vibrate? No, not Mexican jumping beans: These devices usually use a DC motor. DC motors change electrical energy, such as the energy stored in a battery, into motion. That motion may involve turning the wheels of a robot that you build or shaking your pager. In fact, you can use a DC motor in any project where you need motion.

Electromagnets make up an important part of DC motors because these motors consist of, essentially, an electromagnet on an axle rotating between two permanent magnets, as you can see in Figure 5-8.

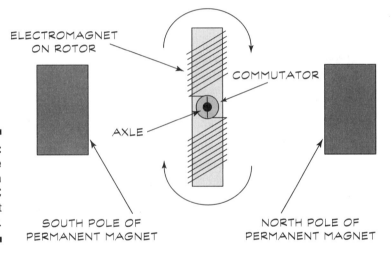

Figure 5-8:
How the parts of a simple DC motor fit together.

ELECTROMAGNET
ON ROTOR

COMMUTATOR

AXLE

SOUTH POLE OF
PERMANENT MAGNET

NORTH POLE OF
PERMANENT MAGNET

The positive and negative terminals of the battery connect so that each end of the electromagnet has the same polarity as the permanent magnet next to it. Like poles of magnets repel each other. This repelling action moves the electromagnet and causes the axle to spin. As the axle spins, the positive and negative connections to the electromagnet swap places, so the magnets continue to push the axle around. A simple mechanism consisting of a *commutator* (a segmented wheel with each segment connected to a different end of the electromagnet) and brushes that touch the commutator cause the connections to change. The commutator turns with the axle and the brushes are stationary, with one brush connected to the positive battery terminal and the other brush to the negative battery terminal. As the axle, and therefore the commutator, rotates, the segment in contact with each brush changes. This in turn changes which end of the electromagnet is connected to negative or positive voltage.

If you want to get a feel for the mechanism inside a DC motor, buy a cheap one for a few dollars and tear it apart.

The axle in a DC motor rotates a few thousand times per minute — a bit fast for most applications. Suppliers sell DC motors with something called a gear head pre-mounted; this device reduces the speed of the output shaft to under a hundred revolutions per minute (rpm). This is similar to the way that changing gears in your car changes the speed of the car.

Suppliers' catalogs typically list several specifications for the motors they carry. Two key things that you need to consider are

- ✔ **Speed:** Listed as rpm (revolutions per minute). The speed that you need depends on your project. For example, when turning the wheels of a model car, you may aim for 60 rpm, with the motor rotating the wheels once per second.

- ✔ **Operating voltage:** The operating voltage is given as a range. Electronics projects typically use a motor that works in the 4.5- to 12-volt range. Also notice the manufacturer's nominal voltage and stated rpm for the motor. The motor runs at this rpm when you supply the nominal voltage. If you supply less than the nominal voltage, the motor runs slower than the stated rpm.

DC motors have two wires (or terminals that you solder wires to), one each for the positive and negative supply voltage. You run the motor by simply supplying a DC voltage that generates the speed that you want and switching off the voltage when you want the motor to stop.

You can use a more efficient method of controlling the speed of the motor called *pulse width modulation.* This method turns voltage on and off in quick pulses. The longer the on intervals, the faster the motor goes. If you're building a kit for something motor-controlled, such as a robot, the electronics for the kit supplies this speed control.

If you're attaching things such as wheels, fan blades, and so on to the motor shaft, be sure that you have attached the component securely before you apply power to the motor. If not, the item may spin off and hit you, or someone near and dear to you, in the face.

So You Want to Make Some Noise?

So, you've probably asked yourself at one time or another, just what is sound? *Sound* is simply a series of vibrations traveling through the air. When you talk, for example, your vocal cords vibrate to create sound waves that travel to a listener's ear.

In electronics, you can use speakers and buzzers to create sound. In fact, the world of electronics tends to be a noisy one: You can activate music, buzzers, alarms, and other sounds with your electronic gadgets. In the following sections, we explore these devices that you can use to get your project heard.

Speaking of speakers

Most speakers simply consist of one permanent magnet, an electromagnet, and a cone. Figure 5-9 shows how the components of a speaker are arranged.

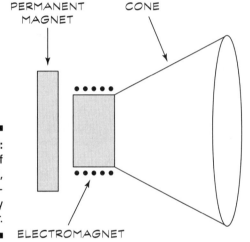

Figure 5-9:
The parts of your typical, garden-variety speaker.

The electromagnet is attached to the cone. When electric current moves through the electromagnet, it either gets pulled toward the permanent magnet or (if the electric current goes in the other direction) pushed away from the permanent magnet. The motion of the electromagnet causes the cone to vibrate, which creates sound waves.

You state the frequency range over which a speaker generates sound in Hz (hertz) or kHz (kilohertz). The human ear can hear sound over a frequency range of about 20Hz to 20kHz. Speakers generate sound over various ranges, depending on their size and design (for example, in a stereo set up, one speaker may generate in the bass range and another in a higher range). If you're just looking for a general-purpose speaker, don't worry too much about the frequency range. If you're building a super-duper high-end audio system, you probably want to spend a lot of time investigating and picking out speakers that meet your needs.

In Chapter 1, we cover the units Hz and kHz in more detail.

Buzzers

Admit it — buzzers are cool. They can do everything from alerting you when somebody's coming into your room to scaring the cat off the couch.

How buzzers work

Here's how a common type of buzzer works: Voltage applied to a piezoelectric crystal causes the crystal to expand or contract. If you attach a diaphragm to the crystal, changing voltage causes the diaphragm to vibrate and generate sound waves. You call these buzzers piezo buzzers, referring to the *piezoelectric effect,* the ability of certain crystals — quartz and topaz to name a few — to expand or contract when you apply voltage to them.

Some buzzers use electromagnets. For beginners, we recommend the piezo buzzer just to minimize the number of moving parts.

Buzzers have two leads and come in a variety of looks. Figure 5-10 shows a couple of typical buzzers. To connect the leads the correct way, remember that the red lead goes to positive DC voltage.

Figure 5-10:
Noisy little
buzzers.

How noisy should your buzzer get?

A buzzer generates sound at one frequency; the specifications for a buzzer indicate several things:

- ✔ **The frequency of sound it emits:** Most buzzers give off sound between 2kHz and 4kHz because humans can hear sound in that range very easily.

- ✔ **The operating voltage and voltage range:** Just make sure that you get a buzzer that works with the DC voltage that your project supplies.

- ✔ **The level of sound it produces in units of decibels (db):** A higher number of db indicates a louder sound. Higher DC voltage (within the voltage operating range of the buzzer) provides higher sound levels.

Be careful that the sound doesn't get so loud that it damages your hearing. You can start to get an annoying ringing in your ears at levels of around 85 db and above.

Part III
Putting It on Paper

In this part . . .

You find your way around the forest of electronics circuits with a schematic diagram, your road map to the components you need and how they connect to each other. In this part, you discover how to read a schematic and how to use a basic schematic to deduce whether a circuit beeps, lights up, spins around, or whatever. After you're done, you'll be able to make sense of what all those squiggly lines actually mean.

Chapter 6

Reading a Schematic

. . . .

. . . .

*I*magine driving cross-country without a roadmap. Chances are, you'd get lost along the way and end up driving in circles. Roadmaps exist to help you find your way. You can use roadmaps for building electronic circuits, as well. They're called *schematic diagrams,* and they show you how all the parts of the circuits are connected. Schematics show these connections with symbols that represent electronic parts and lines that show how you attach the parts.

Although not all electronic circuits that you encounter are described in the form of a schematic, many are. If you're serious at all about studying electronics, sooner or later, you need to understand how to read a schematic. Surprise! The language of schematics isn't all that hard. Most schematic diagrams use only a small handful of symbols for components, such as resistors, capacitors, and transistors.

In this chapter, we tell you all that you really need to know so that you can read almost any schematic diagram you come across.

What's a Schematic, and Why Should I Care?

If you know how to read a roadmap, you're already well on your way to reading a schematic. Schematic diagrams are a lot like maps, where lines connect things. But where a roadmap uses lines to connect dots and stars that represent towns and cities, schematics use lines to connect symbols representing resistors, transistors, and other components that make up a circuit.

Schematics serve two main purposes:

✔ **Show you how to reproduce a circuit.** By reading the symbols and following the interconnections, you can build the circuit shown in the schematic.

✔ **Give you an overview of a circuit so that you can better understand how it works.** You may find this knowledge useful if, for example, you need to repair the circuit or replace a component.

Discovering how to read a schematic is a little like learning a foreign language. On the whole, you find that most schematics follow fairly standard conventions. However, just as you can speak many languages with different dialects, the language of schematics is far from universal. Schematics can vary depending on the age of the diagram, its country of origin, the whim of the circuit designer, and many other factors.

In this book, we use conventions commonly accepted in North America. But to help you deal with the variations that you may encounter, we also show you some other conventions, including those used in Europe, and old-style diagrams that use things like pre-digital age radio tubes.

Getting a Grip on Schematic Symbols

Today's electronic circuit schematics use hundreds of symbols, and older circuits that use tubes and other components common in your grandpa's day use even more symbols. But you're in luck — you need to remember only a few dozen common symbols. The rest you can look up as you go.

In this chapter, we cover the most common electronic symbols, including those for basic components, such as resistors and capacitors; logic symbols, such as the AND and OR gate; transistors; and more. We cover symbols alphabetically within the following four categories:

✔ **Basic schematic symbols:** Chassis and earth ground, connection points, inputs, and outputs

✔ **Electronic components:** Resistors, transistors, diodes, and chokes

✔ **Logic symbols:** AND, OR, NOR, and inverter gates

✔ **Miscellaneous symbols:** switches, light bulbs, and other hardware

Getting the scoop on basic schematic symbols

Your run-of-the-mill, basic schematic symbols represent the mechanical aspects of a circuit, such as the power source, how and where wires are attached, and any connectors, jacks, and terminals.

Take a look at Chapters 4 and 5 for information about common components such as connectors and jacks, and at Chapter 1 for information about electricity basics such as power source and ground.

Power and ground

The symbol for power looks like a long stick with a circle at the top. The symbol for ground is a long line with three horizontal lines at the bottom. Power for a circuit can come from an alternating current (AC) source, such as the 117 VAC outlet in your house or office (so-called "line powered"), or a direct current (DC) source, such as a battery or the low-voltage side of a wall transformer. Ground is a connection used as a reference for all voltages in a circuit.

Here are some of the ways schematics might appear based on the power source that you use:

✔ In line-powered electronic circuits, you typically use an internal power supply to *step down* (or lower) the 117 VAC, and convert it to DC. This lower-voltage DC gets delivered to the components in the circuit. If you're looking at a schematic for a VCR or some other gadget getting its power from a wall outlet, that schematic probably shows both AC and DC power.

✔ In DC circuits, the schematic may have one or more voltage sources, such as +5 VDC, +12 VDC, and even -5 or -12 VDC. If a schematic doesn't specify a voltage, you're often (but not always!) dealing with 5 VDC. And, unless otherwise specifically noted, the voltage in a schematic is almost always DC, not AC.

As we mention in Chapter 1, all electrical connections require a minimum of two wires: one for power and one for ground. You may also hear ground called *return* or *common.* As you can see in Figure 6-1, a schematic may show the ground connection in a number of ways:

✔ **No ground symbol:** The schematic can show two power wires connected to the circuit. In a battery-powered circuit, ground is the negative terminal of the battery.

✔ **Single ground symbol:** The schematic shows all the ground connections connected to a single point. It doesn't show the power source, but the ground always connects to the negative terminal of the battery or DC power source.

✔ **Multiple ground symbols:** In more complex schematics, it is usually easier to draw the circuit with several ground points. In the actual working circuit, all these ground points connect together.

There are two common forms of ground symbols: earth ground and chassis ground (see Figure 6-2). Although schematics often use them interchangeably, these symbols actually mean different things.

An *earth ground* denotes a connection that you attach to the ground wire in your house's electrical system. The third (usually green) wire in a three-wire power cord comes into play as an earth ground, for example.

Conversely, a *chassis ground* is the connection of wires in a low-voltage circuit. The term gets its name because, in older equipment, the metal chassis of the device (hi-fi, television, or whatever) served as the common ground connection. Using a metal chassis for a ground connection is not as common today, but we still use the term "chassis ground," just the same.

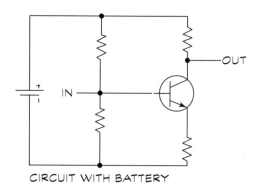

CIRCUIT WITH BATTERY

Figure 6-1:
Ground
symbols can
take many
forms,
including
none
(usually
replaced
with a
battery
symbol),
single, and
multiple.

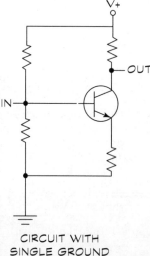

CIRCUIT WITH
SINGLE GROUND

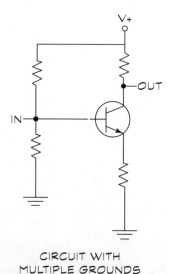

CIRCUIT WITH
MULTIPLE GROUNDS

Figure 6-2:
Earth
ground and
chassis
ground are
different
things,
technically,
and you
represent
them using
specific
schematic
symbols.

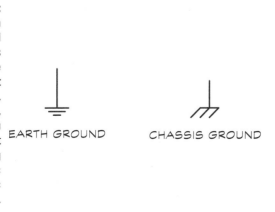

EARTH GROUND   CHASSIS GROUND

 In this book, we use only the schematic symbol for earth ground because most schematics that you see nowadays use that symbol as the standard.

Interconnections

 You connect components in a circuit either by using insulated wires or thin traces of copper on a circuit board. (You can read more about circuit boards and discover how to make your own in Chapter 12.)

Most schematics don't make a distinction about how you connect the components together. That connection is wholly dependent on how you choose to build the circuit. The schematic's representation of the wiring merely shows which wires go where.

Schematics aren't perfect!

On those schematics that use neither a break nor a loop, you only have the presence of the dot to let you know that the wires should connect but no indication when they don't connect. What happens if the person who drew the schematic simply forgot the dot? Sorry to say, you may find errors in schematic diagrams, just like anything else. If you build a circuit from such a schematic, and the circuit doesn't work when you first power it up, you may suspect a missing wire connection somewhere. But unless you're familiar with electronics, determining which connection is missing isn't always easy. In these cases, consult with the person who drew the schematic if you can, just to be sure.

With more complex schematics, some lines may have to cross over others. You need to know when crossed lines represent a wire connection and when they don't. In the ideal modern schematic, you show connecting and non-connecting wires like this:

- ✔ A break or loop indicates wires that don't connect.
- ✔ A dot at the intersection of two lines shows the wires that do connect.

You can see some common variations in Figure 6-3.

This method of showing connections isn't universal, so you have to figure out which wires connect and which don't by checking the drawing style used in the schematic. For example, in any given schematic, connecting wires may simply intersect, without a dot to indicate the intersection, in which case the presence of an intersection is essentially unknown: there might, or might not be a connection.

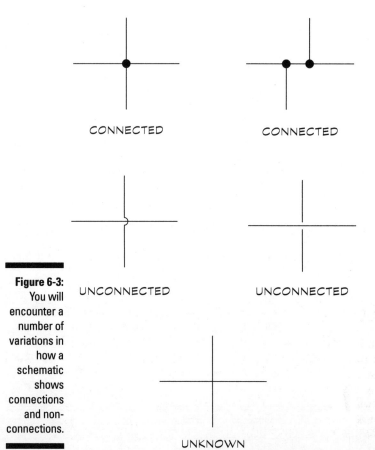

Figure 6-3: You will encounter a number of variations in how a schematic shows connections and non-connections.

Jacks, plugs, and connectors

The first symbol here is a two-wire jack, and the second is a shielded jack, two common types of jacks used in electronics. Most electronic circuits interact with the outside world in some way. Take, for example, a guitar amplifier. At the very least, it has a jack so that you can plug in the guitar's cord. In other circuits that you may encounter, you can use jacks, plugs, or other connectors to interface with things, such as temperature probes, microphones, or battery packs.

Here are the items that you most often use to make these connections:

✔ **Jack and plug:** These two items are a well-matched couple because a *jack* is a receptacle for a *plug*. Or, using these words in a sentence, "Fredricka plugged her headphones into the headphone jack of her shiny new portable CD player."

✔ **Connector:** A *connector* is a generic term for any fitting that allows you to easily connect and disconnect wires to a circuit. The connector may be some elaborate multi-socket arrangement, or it can be just two screw terminals.

Symbols for jacks, plugs, and connectors can vary greatly among schematics. The symbols that we use in this book are among the most commonly used. The exact style of the symbol may vary from one schematic to the next, but the idea still comes across — the connector provides you with a way to attach something to a circuit.

Symbols for electronic components

You can find literally hundreds of symbols for electronic components out there because there are hundreds of them to depict. Fortunately, you probably encounter only a small number of these symbols in the schematics that you run into.

This section begins with a discussion of labels that may accompany any component in a schematic, and then covers the most common component symbols, divided into categories.

As you go through this section, feel free to refer to Chapters 4 and 5 if you need a refresher course on what a particular component does and what you can use it for.

Component symbols like company

Component symbols don't like to travel alone, so each component symbol in a schematic is commonly accompanied by one or more of the following:

✔ **Reference ID:** An identifier, such as "R1" or "Q3." The convention is to use the letter to represent the type of component. The most common letters are R for resistor, C for capacitor, D for diode, L for inductor, T for transformer, Q for transistor, and U or IC for integrated circuit. If you have several items of the same type, the numerical suffix (such as R3) identifies that particular component.

✔ **Part number:** Used if the component is standard, such as a transistor or integrated circuit, or you have a manufacturer's custom product part. For example, a part number may be something like 2N2222 (that's a commonly used transistor) or 555 (a type of integrated circuit used in timing applications).

✔ **A value:** Used if the component doesn't go by a conventional part number. Component values are typically used for non-solid-state parts, such as resistors and capacitors. For example, when indicating a resistor, the value — in ohms, K ohms (thousands of ohms), or megohms (millions of ohms) — could be marked beside the resistor symbol and/or the reference ID.

To indicate the circuit's proper function, the schematic may also include additional specifics about a component. For instance, you commonly can assume that, unless otherwise noted, all resistors are rated at ¼ or ⅛ watt. When the circuit requires a different wattage — say a 1-watt or 10-watt power resistor — you may write that wattage beside the symbol in the schematic. In other cases, you may note any special considerations in the parts list, or as an addendum to the schematic.

Reference ID primer

Components such as capacitors are often identified in a schematic using a letter, such as C for capacitor, followed by the Reference ID number (such as C2). The number identifies a specific capacitor, and can be used in a parts list where you note the precise value of the capacitor to use, if that value isn't also printed beside the capacitor symbol. The following letters are among those most commonly used. Note that some components use abbreviations instead of single letters, but the intent is the same:

C - Capacitor

D - Diode

IC (or U) - Integrated circuit

L - Inductor

LED - Light-emitting diode

Q - Transistor

R - Resistor

RLY - Relay

T - Transformer

XTAL - Crystal

In various headings in this chapter when you see a letter in parenthesis (as in the heading that follows) it's reminding you of the letter abbreviation for that component.

Capacitors (C)

The schematic diagram of a capacitor reflects its internal construction: two conductive plates separated by a small gap. The small gap and its contents are called the dielectric. As we discuss in Chapter 4, the dielectric can be air, liquid, or some form of insulator (such as plastic or mica).

Capacitors can be either polarized or non-polarized. Schematics show the polarity with a + (plus) symbol, though on the capacitor itself, the polarity can be marked with either a + (plus) or – (minus) symbol next to one of the leads.

Crystals and resonators (XTAL)

You use crystals and resonators to provide an accurate time base for electronics circuits. When you use these components with the appropriate oscillator circuit, the crystal or resonator generates a series of pulses, sort of like a metronome. The symbol for a crystal looks a lot like a capacitor, except that a crystal symbol has a rectangle between the end plates.

Diodes (D)

You can find many kinds of diodes out there, including rectifiers, Zeners, and light-emitting diodes (LED). Figure 6-4 shows an assortment of the most common diode types: The standard rectifying diode, a Zener, the LED, and the photodiode. You can use LEDs as indicator lights and photodiodes to detect light. The sensor for your VCR's remote control is an example of a photodiode. And you commonly find bridge diodes in power supply circuits that convert AC voltage to DC.

STANDARD DIODE   ZENER DIODE

Figure 6-4:
Symbols for
different
types of
diodes.

LED   PHOTODIODE

Inductors (L)

Inductors are coils of wire that you often see used in radio frequency (RF) circuits, such as AM radios and transmitters. The symbols for various types of inductors are quite similar to each other and easy to spot; the main difference among inductors relates to what makes up the core. The most common core materials are air and iron.

Operational amplifiers (U or 1C)

Operational amplifiers are a type of integrated circuit because they are actually circuits within a circuit. They combine, in a single component, all the necessary circuitry to amplify a signal. Schematics commonly use the symbol that you see here for almost any amplifier, not just an operational amplifier. The basic operational amplifier (op-amp) has two inputs (one shown with a + sign, the other with a – sign) and a single output.

Relays (RLY)

You use relays to open and close a circuit while using another voltage (typically a smaller one) as a control. Relays differ from one another in the number of contacts that they contain. The symbol you see here is a double-pole, single-throw (DPST) relay. When working with relays, be sure to keep the control voltage (shown connected to the coil) separate from the contact voltage (shown connected to the contacts of the relay) because the two voltages may differ and are not intended to be switched.

Resistors (R)

Resistors may be the most common component of any electronic circuit. Resistors can be either fixed or variable. In a fixed resistor, the resistance never varies. In a variable resistor, the resistance can be changed. What effects the change depends on the construction of the variable resistor. In some cases you manually effect the change, for example, by turning a knob; or the change can be caused by an outside stimulus, such as a change in light, voltage, or temperature. See the section "One Size Fits All: Adjustable Components," later in this chapter for more about variable resistors.

Transistors (Q)

You often use a transistor in circuits to function either as a switch or as an amplifier. Most transistors have three wires (sometimes four). The arrows in the symbol indicates the type of transistor. For example, in a bipolar PNP transistor type, the arrow faces the base. In a bipolar NPN transistor type, the arrow faces away from the base. (To catch up on the parts of a transistor, such as the base, you can review Chapter 4).

Bipolar transistors are among the most common transistors, but you can also run into other transistor types, such as the field-effect transistor (FET) and the unijunction transistor (UJT). Also note that there are light-sensitive transistors that switch on when exposed to light. You can see symbols for PNP, NPN, and FET transistors types in Figure 6-5.

Transformers (T)

Transformers do just what their name implies: They transform an electric current and voltage to either a higher or lower value. You commonly find transformers in two sections of a circuit:

✔ **Power supply section:** Where you use the transformer to step down the 117 VAC line voltage to a lower level, such as 12 or 18 volts

✔ **Audio output section:** To change the *impedance* (the measure of opposition in an electrical circuit to a flow of alternating current) of the circuit to a level suitable for driving an audio speaker

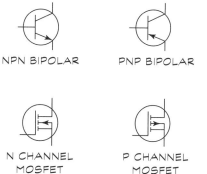

Figure 6-5:
Variations on a theme. These symbols represent various types of transistors.

NPN BIPOLAR

PNP BIPOLAR

N CHANNEL MOSFET

P CHANNEL MOSFET

Logic gate symbols

Schematic diagrams for many digital circuits use logic gate symbols. Logic gate symbols indicate the action that occurs in response to the two possible voltage states (on or off). Other than power, these voltage states are the only two present in a digital circuit. You can see the most common logic symbols in Table 6-1.

REMEMBER

If logic gates are a mystery to you, go to Chapter 5 to study up on what they are, how they work, and their possible states of on/off or high/low.

Table 6-1	Common Logic Gate Symbols	
Name	*Symbol*	*Function*
AND		Output is binary 1 only if both inputs are binary 1
NAND		Same as AND, but output is inverted (binary 0)
OR		Output is binary 1 if either input is binary 1

(continued)

Table 6-1 *(continued)*

Name	Symbol	Function
NOR		Same as OR, but output is inverted (binary 0)
Buffer		Provides a protective buffer or additional drive current between two circuits
Inverter		Similar to Buffer, but the output is inverted
Flip flop		Output toggles between 0 and 1

Although you can create AND, OR, and other digital logic gates with transistors, most circuits use an integrated circuit chip (called an IC). One IC contains a number of individual logic gates. For example, the 7400 integrated circuit contains four gates sharing a single power connection.

Some schematics show individual logic gates, and some show connections to the full integrated circuit. You can see an example of each in Figure 6-6. Whether the schematic uses individual gates or an entire IC package, it usually notes the power connections. When it doesn't, you have no choice but to look up the so-called pinout of the device in a reference book. The *pinout* is a reference sheet that indicates what each of the connections, or pins, of the integrated circuit is used for. You can often find pinout diagrams on data sheets that manufacturers of integrated circuits provide. You can locate them on the Web using your favorite search engine.

Miscellaneous symbols

You may run across several miscellaneous symbols used in schematics to represent various kinds of electronic gear. For the most part, these symbols are self-explanatory, so we keep things simple and to the point in this section.

However, take special note of the symbols used for switches. The schematic symbol for the switch indicates the number of *poles* (connections) and positions in the switch. Each pole can switch a different part of the circuit, such as a portion of a circuit that requires a different voltage. (Switches are covered in more detail in Chapter 7).

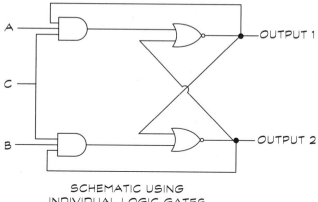

SCHEMATIC USING
INDIVIDUAL LOGIC GATES

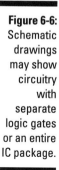

Figure 6-6:
Schematic
drawings
may show
circuitry
with
separate
logic gates
or an entire
IC package.

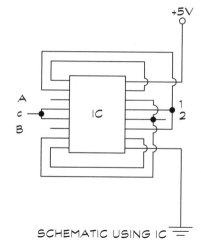

SCHEMATIC USING IC

Here are some common switch types, and some variations you will encounter
as you build electronics projects:

✔ A single-pole, single-throw (SPST) switch has one position (on-off) or
"throw," and only one pole.

✔ A double-pole, double-throw (DPDT) switch has two positions (on-on, or
on-off-on) and two poles.

✔ Other variations include DPST (double-pole, single-throw) and switches
with three or more poles.

✔ In addition to the poles and throws, some switches are spring-loaded
(called "momentary switches"). These switches are either normally
open (NO) or normally closed (NC). The normal state occurs when the
switch isn't being pressed. For example, a normally open switch doesn't
make electrical contact until you press it.

Table 6-2 shows several common switch symbols, such as SPST and DPDT, along with a couple of other symbols for such components as speakers, batteries, and incandescent lamps.

Table 6-2	Miscellaneous Component Symbols
Name	*Symbol*
Switch, SPST	
Switch, SPDT	
Switch, DPDT	
Switch, normally open	
Switch, normally closed	
Speaker	
Piezoelectric buzzer	
Battery	
Meter	
Incandescent lamp	

Getting Component Polarity Right

Many, though not all, components use polarized connections. To make sure that these components function properly, you have to connect them in the circuit in just the right way. In some instances, reversing a component from its proper polarity can permanently damage it and other components in the circuit.

Figure 6-7 shows how schematics identify polarity when using various components.

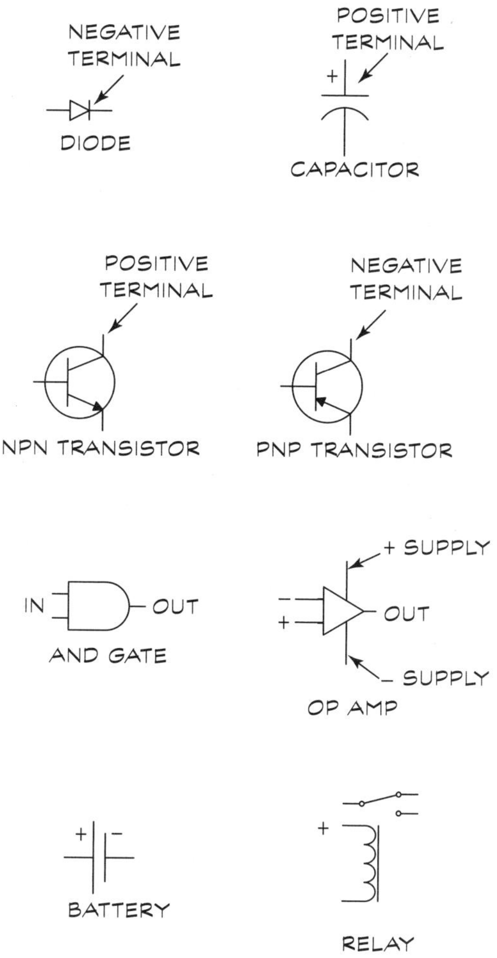

NEGATIVE
TERMINAL

DIODE

POSITIVE
TERMINAL

+

CAPACITOR

POSITIVE
TERMINAL

NEGATIVE
TERMINAL

NPN TRANSISTOR

PNP TRANSISTOR

IN — OUT

AND GATE

+ SUPPLY

—
+ — OUT

— SUPPLY

OP AMP

Figure 6-7:
Polarity
symbols for
several
components.

+ —

BATTERY

+

RELAY

Be sure to observe polarity when working with the following common
components:

✔ **Diodes:** Including rectifier, Zener, and light-emitting types. Schematics
indicate polarity with a short line, which represents the *cathode* (negative) terminal of the diode.

✔ **Some capacitors:** Electrolytic, tantalum, and several other special types.
Schematics indicate polarity with a + (plus) sign.

✔ **Transistors:** Schematics show polarity by the type of the symbol.

✔ **Logic gates and other integrated circuits:** Schematics show polarity by
the labels or other markings on the symbol.

✔ **Op amps:** An op-amp has three connections (besides power): two inputs
and the output. The inputs are marked + (non-inverting) and (–) inverting.

   ✔ **Battery:** Schematics show polarity with a + (plus) and sometimes also – (negative) sign.

   ✔ **Relay:** Coil only. Schematics show polarity with a + (plus) sign.

One Size Fits All: Adjustable Components

Several types of electronic components are adjustable. Instead of operating at just one value, you can manually adjust the component to operate at a range of values.

The most common adjustable components you encounter (whose symbols you can see in Figure 6-8) are

   ✔ **Variable resistor:** Also called a potentiometer (or pot). Perhaps the most common of all variable components, you use this resistor for volume control, dimming lights, and thousands of other applications. The potentiometer consists of a resistive element wired between two terminals (such the filament in a lightbulb). On a third terminal, a wiper registers changing resistance as you turn the potentiometer knob. Figure 6-8 shows how schematics present potentiometers.

   ✔ **Variable capacitor:** You most often use a variable capacitor in a tuning circuit, such as an AM radio. The capacitor consists of two or more metal plates separated by air. Turning the knob changes the capacitance of the device.

   ✔ **Variable coil:** Like a variable capacitor, you most often use a variable coil in a tuning circuit. A typical construction uses a coil of wire surrounding a movable metal slug. By moving the slug, you change the inductance of the coil.

Figure 6-8:
Schematic symbols for variable components use an arrow or other mark to show that the value of the component is changeable.

VARIABLE RESISTOR   VARIABLE CAPACITOR   VARIABLE COIL

For a refresher course on the concepts of capacitance and inductance, see Chapters 4 and 5.

Photo-Sensitive Components Help You See the Light

Special light-dependent versions of resistors, diodes, and transistors react to changes in illumination. The value of the component varies depending on the amount of light that strikes it. Most schematics show the light-sensitive nature of the component by using one or two arrows pointing into the body of the component.

Table 6-3 shows several common schematic symbols for light-sensitive components: Photocells/photoresistors, photodiodes, phototransistors, and solar cells.

Table 6-3	Photo Sensitive Component Symbols	
Name	*Symbol*	*Function*
Photocell/Photoresistor*		Light-sensitive version of a resistor
Photodiode		Light-sensitive version of a diode
Phototransistor		Light-sensitive version of a transistor
Solar cell		Generates electricity in response to sunlight

* Note that you can use the terms photocell and photoresistor interchangeably.

Alternative Schematic Drawing Styles

The schematic symbols in this chapter belong to the drawing style used in North America (particularly in the United States) and in Japan. In some countries, notably European nations as well as Australia, somewhat different schematic symbols are used. If you're using a schematic for a circuit not designed in the United States or Japan, you need to do a wee bit o' schematic translation in order to understand all the components.

Figure 6-9 shows a sampling of schematic symbols commonly used in the United Kingdom and Europe. Notice that there are some obvious differences in the resistor symbols, both fixed and variable.

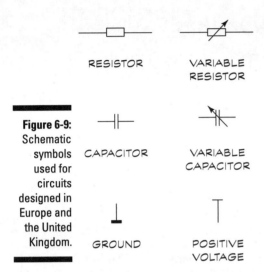

RESISTOR   VARIABLE RESISTOR

Figure 6-9: Schematic symbols used for circuits designed in Europe and the United Kingdom.

CAPACITOR   VARIABLE CAPACITOR

GROUND   POSITIVE VOLTAGE

This style organizes its symbols differently than the American style. In the United States, you express resistor values over 1,000 ohms in the form of 6.8K or 10.2K, with the *K* following the value. The European schematic style eliminates the decimal point. Typical of schematics you'd find in the United Kingdom are resistor values expressed in the form of 6K8 or 10K2. This style substitutes the K (which stands for kilohms, or thousands of ohms) for the decimal point.

You may encounter a few other variations in schematic drawing styles, but all are fairly self-explanatory and the differences are not substantial. After you learn how to use one style of drawing, the others come easily.

Chapter 7

Understanding the Basics of Electronics Circuits

*I*magine that you're building a cute little cottage rather than an electronic gadget. You have to know about the tools and materials that you need to build the thing, and you need to gain skills, such as carpentry and plumbing. But before you begin sawing and plumbing, you have to have a blueprint that gives you an idea of what the final product should look like. That's what a *schematic* is: A blueprint of an electronic circuit that forms the basis of your electronic gadget.

This chapter covers the basics of electronic circuits and examines the basic building blocks that let you trace through the schematic for any project and understand how a circuit functions. *There's one prerequisite with this chapter: it's really important that you read Chapter 6 before you read this one so you don't get lost.*

What the Heck Is a Circuit?

An electronic *circuit* is simply a collection of components connected together with wires through which an electric current moves. You can think of a circuit as composed of five parts:

- ✔ A power source
- ✔ Components, such as resistors and transistors
- ✔ Wires to connect everything together
- ✔ An output device (also referred to as the load), such as a speaker
- ✔ Ground to complete the circuit
- ✔ Many, but not all circuits, also have an input

A Very Basic Circuit

You don't want to start building houses by tackling a 36-room mansion with a complex home stereo system wired into the walls and a maze-like set of secret passages in the basement, right? You also shouldn't start your exploration of electronic circuits with anything overwhelming. So we start you off with the equivalent of building a shed: A simple circuit that powers a light bulb.

Powering a light bulb

One of the simplest circuits that you encounter involves a light bulb and two wires that connect the bulb to a power source. However, you may not find this circuit very practical because the light bulb is always on. Adding a switch to turn the light on or off makes the circuit much more useful. Figure 7-1 shows the schematic of a circuit that contains a light bulb and switch.

Figure 7-1:
This circuit
powers a
light to
chase away
the dark.

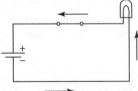

The circuit in Figure 7-1 has the switch in the closed position. When it's closed, it completes the circuit and allows electrons to travel from the negative battery terminal through the light bulb to the positive battery terminal. The light bulb contains a filament that heats up and emits light when the electrons pass through it.

On the other hand, when you have the switch in the open position, such as shown in Figure 7-2, there's a break in the circuit. Because of this break in the circuit, electric current can't flow. No current, no light.

Figure 7-2:
The circuit with an open switch puts you in the dark.

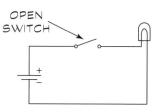

A flashlight works in the same way. When you turn on the flashlight, a switch completes the circuit between the light bulb and the battery and allows electric current to flow. When you turn off the flashlight, you open the circuit, which prevents electric current from flowing.

Controlling the current with a resistor

Say you're building a model railroad and you want to dim the light over the station platform. Just add a resistor to the circuit. Figure 7-3 shows the circuit in Figure 7-1 with a resistor added.

Figure 7-3:
Adding a resistor allows you to dim the light.

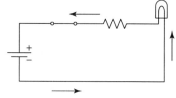

In Chapter 4, we explain that *resistors* "resist" electric current (makes sense, huh?). Adding a resistor reduces the amount of electrons flowing through the circuit. When fewer electrons flow through the filament in the light bulb, the filament emits less light.

You can use Ohm's Law (for a quick review of this handy rule, see Chapter 1) to calculate the amount of current flowing through this circuit before and after you add the resistor. If the resistance of the light bulb is 5 ohms and the battery applies 3 volts, then you calculate the current like this:

$$I = \frac{V}{R} = \frac{3 \text{ volts}}{5 \text{ ohms}} = 0.6 \text{ amp}$$

Here I represents the current, V stands for the voltage, and R represents the resistance.

When you add a 5-ohm resistor to the circuit, the total resistance of the circuit becomes 10 ohms, and you calculate the current as:

$$I = \frac{V}{R} = \frac{3 \text{ volts}}{10 \text{ ohms}} = 0.3 \text{ amp}$$

The resistor cuts the current running through the light bulb's filament in half. This current cutting reduces the amount of light over your train station platform, allowing the tiny stationmaster to catch a few winks.

Parallel (or Series) Parking Your Light Bulbs

You can arrange components in series so that the same current runs through each component, or you can arrange them in parallel so that one batch of current runs through one component and another batch of current goes through another component, and so on. In the following sections, you can see just how series and parallel circuits work.

Circuits: The series

In the circuit in Figure 7-3, electrons flow from the negative battery terminal, through the light bulb and then go on to run through the resistor before reaching the positive battery terminal. You call this set-up a *series circuit,* meaning that the current runs through each component sequentially. You can calculate the total resistance of a series circuit simply by adding together the resistances of each component.

Figure 7-4 shows another example of a series circuit with 4 resistors.

To calculate the total resistance of this circuit, or R_t, simply add the values of all 4 resistors:

$$R_t = 220 \text{ } \Omega + 33 \text{ } \Omega + 10 \text{ } \Omega + 330 \text{ } \Omega = 593 \text{ } \Omega$$

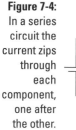
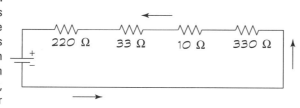

Figure 7-4:
In a series circuit the current zips through each component, one after the other.

You can use this value of R_t with Ohm's Law to calculate the current in the circuit. So, if +V (supply voltage) equals 9 volts:

$$I = \frac{V}{R} = \frac{9 \text{ volts}}{593\,\Omega} = 0.015 \text{ amps or } 15 \text{ milliamps}$$

Why should I care about the total current in a circuit, you ask? There are two really good reasons:

- ✔ Even the hardiest components can only handle a certain amount of current; for example, an LED would probably burn up if you ran more than 50 milliamps through it.

- ✔ On the other hand, your power supply or batteries can only supply a given amount of current. The level of current calculated here, 15 milliamps, is no big deal. However, the next example uses over 1 amp of current, which raises the bar for your power supply or battery. *Bottom line:* To make things run, make sure that you have an adequate power source to supply as much current as the circuit requires for as long as you need it to run.

TIP

There is a potential problem that you may run into with series circuits: If one component fails, it stops the flow of current to every component in the circuit. So, if your spiffy new restaurant sign sports 200 light bulbs wired together in series and one burns out, every one of the light bulbs goes dark.

Parallel circuits

There's a way to fix the problem of all components in a series circuit blacking out when one item fails. You can wire components in a parallel circuit, such as the circuit in Figure 7-5. With a parallel circuit, if you burn out a few bulbs in your restaurant sign, the rest of it stays lit. (Of course, you may be left with a glowing sign reading, "WORLD'S BEST FOO." There are pros and cons to everything.)

Figure 7-5:
A parallel
circuit won't
fail if a
single
component
burns out.

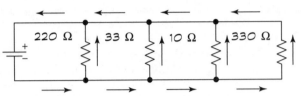

Here's how the parallel circuit in Figure 7-5 works: Electrons flow from the negative battery terminal, through each resistor, and finally to the positive battery terminal. The electrons flowing through one resistor don't flow through the other resistors. So, if your restaurant sign has 200 light bulbs wired together in parallel and one burns out, light still shines from 199 light bulbs.

You calculate the total resistance of the circuit in Figure 7-5, referred to as R_t, by using the following equation:

$$R_t = \frac{1}{\frac{1}{220\,\Omega} + \frac{1}{33\,\Omega} + \frac{1}{10\,\Omega} + \frac{1}{330\,\Omega}} = 7.2\,\Omega$$

In a series setup, you calculate R_t by finding the sum of all the resistances. In a parallel circuit, the R_t of the circuit is a smaller value than the smallest resistor (in Figure 7-5, 7.2 Ω versus 10 Ω for the smallest resistor).

You can calculate the total current running through this circuit by using R_t in Ohm's Law. Again, using a +V of 9 volts, you get a total of 1.25 amps using this equation:

$$I = \frac{V}{R_t} = \frac{9 \text{ volts}}{7.2\,\Omega} = 1.25 \text{ amps}$$

In this example, if you run your project off of batteries, you probably drain them of power in a relatively short time. Batteries have ratings of amp hours. A battery with a rating of one amp hour only lasts for an hour with a circuit drawing one amp. Therefore, your decision about what power source to use must take into account both the current a circuit draws and how long you want to run the circuit.

Exploring a Voltage Divider Circuit

Time for a test (don't worry, its open book): In Chapter 1, we state that *voltage* is a force that pulls electrons through a wire. If you put that together with the information we provide in Chapter 4, that *resistors* "resist" electrons going

through them, what can you conclude? If you said that as voltage pulls electrons through resistors (or any other component), the resistor uses up some
of the voltage, you'd get an A+. You call this lowering of voltage a *voltage drop.*

A circuit called a *voltage divider* uses voltage drops to produce voltage lower
than the supply voltage at specific points in the circuit. Figure 7-6 shows a
voltage divider circuit. For example, it's standard to supply 5 volts to a transistor, but let's say you have a power supply of 9 volts. You can use a voltage
divider to reduce the voltage to 5 volts.

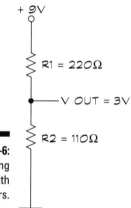

+ 9V

R1 = 220Ω

V OUT = 3V

R2 = 110Ω

Figure 7-6:
Dividing
voltage with
resistors.

The voltage dropped across each resistor is proportional to the value of the
resistor divided by the total resistance, like this:

$$\text{Voltage dropped across} R1 = \frac{R1}{R1 + R2} \times V_t = \frac{220\,\Omega}{330\,\Omega} \times 9\,V = 6\,V$$

You calculate the output voltage by taking the supply voltage minus the voltage dropped across the resistor, R1:

$$V_{out} = V_{in} - V_{R1} = 9 \text{ volts} - 6 \text{ volts} = 3 \text{ volts}$$

But what if you need a different output voltage? Simply change the resistors.
For example, if you want the output voltage to be half of the supply voltage,
just use two resistors that have the same value. Then using the equation to
calculate the voltage dropped across R1:

$$\text{Voltage dropped across R1} = \text{one half of } V_t.$$

If you need the output voltage to be two-thirds of the supply voltage, use an
R1 that is one half the resistance of R2. Then using the equation to calculate
the voltage dropped across R:

$$\text{Voltage dropped across R1} = \text{one third of } V_t.$$

Measuring Current with Voltage

Just as the average human body can take only so much fast food, the most common testing tool of electronics, multimeters, can handle only so much current. When the current flowing through a circuit is too high for you to measure directly with your multimeter (see Chapter 9 for more about working with multimeters), you can measure the voltage drop across a resistor with the multimeter instead and calculate the current from the voltage drop. Figure 7-7 shows a sample circuit where a very small value resistor has been inserted into the circuit to allow you to make this measurement without disturbing the values in the circuit.

Figure 7-7: Just adding one little resistor allows you to measure voltage to figure out the current.

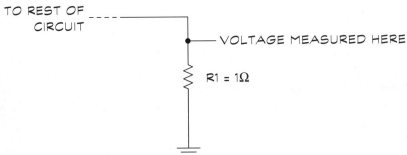

TO REST OF CIRCUIT

VOLTAGE MEASURED HERE

R1 = 1Ω

In this example, you place the resistor in an existing circuit in series with the other components to determine the amount of current flowing in the circuit. You use a 1-ohm, 10-watt resistor because you don't need to worry about a change of 1 ohm resistance in most circuits; the 10-watt rating prevents the resistor from being burned up in most circuits.

Use the multimeter to measure the voltage drop across the resistor, from the voltage measurement point (noted in Figure 7-7) to ground. You can then use Ohm's Law to calculate the current. If, for example, the multimeter measures 2 volts, you calculate the current in this way:

$$\text{Current} = \frac{V}{R} = \frac{2 \text{ volts}}{1\Omega} = 2 \text{ amps}$$

You probably should check the power that you plan to run through the resistor to ensure that the resistor doesn't burn up like Atlanta in *Gone with the Wind*. Calculate the power the resistor will draw by using another form of Ohm's Law:

$$\text{Power} = R \times I^2 = 1 \ \Omega \times (2 \text{ amps})^2 = 4 \text{ watts}$$

Using this equation you know roughly how much power the resistor will draw based on your estimate of the amount of current in the circuit. Try to stay 25% below the power rating of the resistor or it could get REALLY hot.

In most cases, a 10-watt resistor can withstand the demands of a simple electronics project. If you're burning out 10-watt resistors right and left, you've moved beyond electronics hobbyist to master electrician, and you need to buy a much more advanced electronics book.

What a Team: Capacitors and Resistors

Batman and Robin. Butch Cassidy and the Sundance Kid. Capacitors and resistors . . . Huh? It's true: Capacitors and resistors often team up in an electronic circuit. In fact, a capacitor and resistor arranged in a circuit make up one of the basic building blocks of electronic circuits, such as the one shown in Figure 7-8.

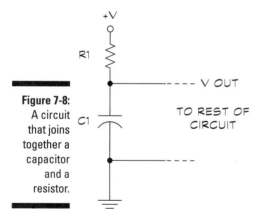

Figure 7-8:
A circuit that joins together a capacitor and a resistor.

So why do these two make such a great team? That's what this section is all about.

How the dynamic duo of resistors and capacitors works

A *capacitor* stores electrons, and a *resistor* controls the flow of electrons. Put these two together, and you can control how fast electrons fill (or charge) a capacitor and how fast those electrons empty out (or discharge) from a capacitor.

The larger the value of a resistor, the less current flows through it for a given voltage, which means it takes more time to fill a capacitor. Likewise, larger capacitors require more electrons to fill them up, which means they take a longer time to charge. By picking the combination of capacitors and resistors, you can determine your project's charge or discharge time.

Turning things on and off

It turns out that the voltage out (V_{out}) depends on how full the capacitor in your circuit is. The closer to full, the higher V_{out}. The closer to empty, the lower V_{out}. Because components use different levels of V_{out}, you can pick values of resistance and capacitance to turn circuits on and off at a certain frequency or after a certain amount of time.

What if you want a capacitor to charge in 30 seconds? You have a 15-microfarad capacitor handy in your parts bin (well, who doesn't?); using a 2-megohm resistor sets the time that it takes the capacitor to get to two-thirds of its capacity.

Filling the capacitor to two-thirds of its capacity often gives a high enough V_{out} to turn on the next component in the circuit. If it doesn't, try a smaller resistor so that the capacitor fills up faster. You can generally do things that simply; take a capacitor that you have handy and calculate how many ohms you need to get close to the desired seconds of delay.

You can calculate the time to fill a capacitor to two-thirds of its capacity using something called an *RC time constant*. Simply multiply the values of the resistor, in ohms, by the capacitor, in farads, and you get the time it takes to fill the capacitor up to two-thirds of its capacity. (In Chapter 1, we discuss how to change 15 microfarads to 0.000015 farads, a procedure we follow in the equation below).

RC time constant = R x C = 2,000,000 ohms x 0.000015 farads = 30 seconds

Giving voltage fluctuations the boot with capacitors

You can use the ability of capacitors to gather and release electrons to smooth out voltage fluctuations. A given voltage level across a capacitor produces a certain number of stored electrons. When the voltage starts to rise, the capacitor stores more electrons, which dampens any rise in voltage. When the voltage drops, the capacitor releases some of its trapped electrons, which dampens the drop in voltage. Power supplies that convert AC to DC often use capacitors to smooth out fluctuations in voltage.

If you want to fine-tune the delay, use a resistor with a slightly smaller value than you need and add a potentiometer (a variable resistor that allows for continual adjustment of resistance from virtually no ohms to some maximum value) in series with the resistor. Because the total resistance is the sum of the value of the resistor and the potentiometer, you can increase or decrease the resistance by adjusting the potentiometer. Just tweak the potentiometer until you get the delay you want. Note that we cover potentiometers in more detail in Chapter 4.

Talking of Transistors

The word transistor doesn't come from some obscure Latin noun; actually the man who built the first one, Walter Brattain, figured that, just as the vacuum tube had the property of transconductance, this new thingie had the electrical property of transresistance. He also knew that a number of electronic devices had come out recently with names, such as varistor and thermistor. Transistor seemed to fit the bill, which is all well and good, but what exactly is the thing? Simply put, a *transistor* controls the flow of electric current by opening and closing a kind of valve within it.

You can use transistors as either a switch or an amplifier. In the following sections, we describe both applications.

Using a transistor as a switch

A switch simply opens or closes a path through which current flows. You can use a transistor as an electrically operated switch. You can see the circuit for a transistor used as a switch in Figure 7-9.

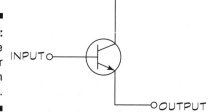

Figure 7-9: You can use a transistor to switch on current.

Take a closer look at what makes up a transistor. A transistor has three leads: Base, emitter, and collector (which we discuss in Chapter 4). When you use a transistor as a switch, the base lead of the transistor works like the toggle on a mechanical switch.

When you're not applying current to the base (that is, there's no input current), the transistor is off, which is equivalent to an open switch. Even with a voltage difference between the other two leads of the transistor, no current flows through the transistor.

When you supply current to the base of the transistor, it turns the transistor on, which is equivalent to a closed switch. With the transistor on, a voltage difference between the other two leads of the transistor causes a current to flow through the transistor and out to whatever doohickey you want to turn on.

How does this on-off thing work in practice? Say that you use an electronic gadget to automatically scatter chicken feed at dawn. The gadget is controlled by a photodiode (similar to a solar cell) in your henhouse, which supplies the input to the transistor. At night, the photodiode doesn't generate any current, and the transistor is off. When the sun rises, the photodiode generates current, and the transistor turns on. When the transistor turns on, current goes to the gizmo that you built to scatter chicken feed so that you can sleep late and the chickens stay happy.

Wait a minute you ask, why not just supply the current from the photodiode to the gizmo? Your gizmo might need a larger current than can be supplied by the photodiode. For example, it might need the current you get from a battery. By using the transistor as a switch you can control the current from the battery with the much smaller current supplied by the photodiode.

In ICs (integrated circuits) that contain logic gates like the ICs used by calculators and computers, transistors wired as switches are an integral part.

When is a transistor an amplifier?

We all need a helping hand from time to time. Why should electronic signals be any different? You often need to amplify signals to get things done. For example, you may have to amplify a signal from a microphone to drive a speaker. Figure 7-10 shows the circuit of a basic one-transistor amplifier.

An amplifier must have a transistor partially turned on. To turn the transistor partially on, you apply a small voltage to the base of the transistor. This procedure is called *biasing* the transistor. In the example in Figure 7-10, in order to bias the transistor, resistors R1 and R2 are connected to the base of the transistor and configured as a voltage divider (see the section "Exploring a Voltage Divider Circuit," earlier in this chapter). The output of this voltage divider supplies enough voltage to the base of the transistor to turn the transistor on and allow current to flow through it.

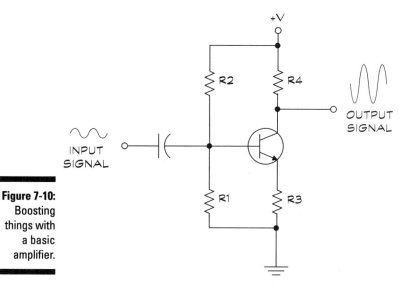

When the amplifier receives an AC input signal, such as from a microphone, the signal must be centered around 0 (zero) volts to maintain the bias. The capacitor at the input filters out any offsets from 0 (zero) volts DC (called a DC offset) in the input signal. You can see this effect in Figure 7-11.

This biased state is the major difference between using a transistor as an amplifier and using a transistor as a switch. When you use a transistor as a switch, you have the transistor either off or on. When using a transistor as an amplifier, you apply a voltage, or bias, to the base to keep the transistor partially turned on. Think of it like keeping a car running at idle.

Figure 7-11:
Filtering out
DC offsets
maintains
the bias.

There's an advantage in leaving a transistor biased on because it responds to any change in the input signal. A transistor requires about 0.6 volts applied to the base (from the base to the emitter) to turn on. If you don't have the transistor turned on, any input signal below 0.6 volts doesn't produce an output signal. With the transistor biased on, it amplifies the entire input signal. Figure 7-12 shows the effect of biasing a transistor on the output signal. Note that in the output signal without bias, only a portion is amplified; the rest is lost. In the output signal with bias, the entire signal is amplified.

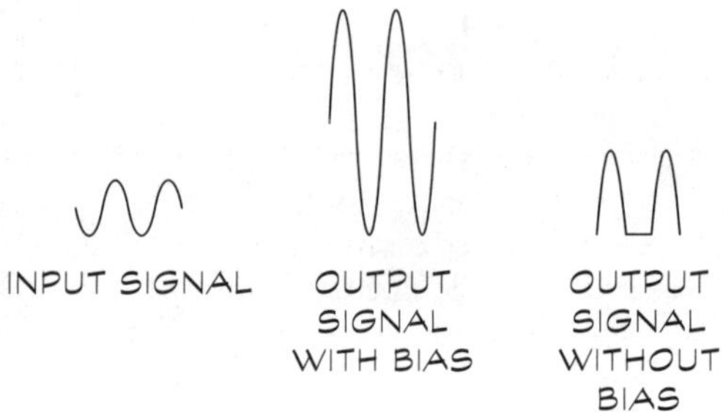

Figure 7-12:
The middle
output
signal has
been
amplified all
the way.

INPUT SIGNAL

OUTPUT
SIGNAL
WITH BIAS

OUTPUT
SIGNAL
WITHOUT
BIAS

The other two resistors that you can see in the circuit in Figure 7-10, R4 between the emitter and ground and R3 between the collector and +V, control the gain. The *gain* is simply how much the signal is amplified. For example, with a gain of 10, a 1-volt input signal becomes a 10-volt output signal.

What else can you do with transistors?

The circuit that we discuss in this section is a common emitter circuit. You can also do the following things with circuits that include transistors:

✔ Wire them in common base circuits, which you use in radio frequency applications or voltage regulators.

✔ Use PNP transistors rather than the more commonly used NPN transistors.

✔ Wire them with more than one transistor, producing multiple stages of amplification.

Let's have more transistor amplifiers

The section titled "Talking of Transistors" gives you a taste of transistor amplifiers, but where's the rest of the meal, you ask? In this chapter, we explain the basics of electronics circuits; we don't really have room to give transistor amplifiers a thorough going-over. If you now have a basic understanding of how transistor amplifiers work and can look at the schematic for a project and understand how the transistors are being used, we're happy.

For you budding electronics Einsteins who just have to know more, try getting your hands on a good electronics design book, such as *The Art of Electronics* by Thomas C. Hayes and Paul Horowitz (Cambridge University Press). It's not cheap, but it's a classic.

An Operational Amplifier

If one transistor is good, more transistors are better, right? An *operational amplifier* is an IC containing several transistors, as well as other components. An operational amplifier, usually called simply an *op amp,* performs much better than an amplifier made from a single transistor. For example, an op amp can provide uniform amplification over a much wider range of frequencies than can a single-transistor amplifier.

Check out a basic circuit that uses an op amp in Figure 7-13.

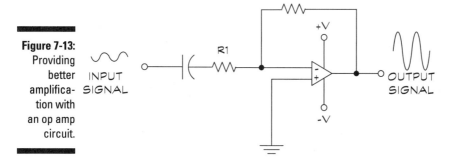

Figure 7-13:
Providing
better
amplifica-
tion with
an op amp
circuit.

Just put a signal (for example, from a microphone) to the input; the signal, amplified several times, then appears at the output, where it can drive a component, such as a speaker. The values of the resistors adjust the gain of the amplifier (remember, gain simply means how much the signal is amplified). You calculate the gain by dividing R2 by R1:

$$\text{Gain} = \frac{R2}{R1}$$

If R2 is 10 times R1, the gain is 10. This gain results in a 1-volt input signal producing a 10-volt output signal.

An op amp requires both negative and positive supply voltages. A positive supply voltage in the range of 8 to 12 volts and a negative supply voltage in the range of -8 to -12 works.

The circuit in Figure 7-13 uses the op amp in an *inverting mode,* which means that the input signal is flipped to produce the output signal. You generally should use the inverting mode because of signal noise problems that you can encounter with the non-inverting mode.

Simplifying a Project with an Integrated Circuit

Whoever said that less is more must have been a fan of integrated circuits (ICs). Using an IC in a project allows you to substitute one component for several because many components are built into an IC. In this section, you discover how to connect an IC into a circuit by connecting inputs, outputs, ground, power, and some resistors and capacitors to the correct pins of the IC, as you can see in Figure 7-14.

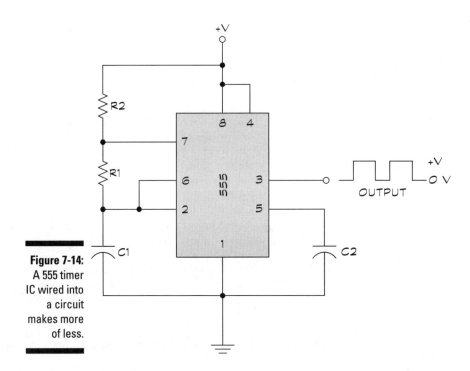

Figure 7-14: A 555 timer IC wired into a circuit makes more of less.

Which pin number you use to connect different parts of the circuit depends on the design of the IC. You can identify these pins for each IC on the manufacturer's data sheet or the schematic for a particular project.

In Figure 7-14:

- ✔ +V connects to pin 8, which is power, and pin 4, which is reset.

- ✔ Ground connects to pin 1.

- ✔ The output of the IC is at pin 3.

✔ Pins 2 and 6, referred to respectively as the trigger and the threshold, connect to the circuit between the capacitor and resistor, R1.

✔ Pin 7, the discharge pin, connects to the circuit between R1 and R2.

When you connect a 555 timer IC (an integrated circuit as a timer) in this way, it generates a digital waveform from the output. The frequency of the waveform depends on how fast the capacitor fills and drains. You calculate how fast the capacitor fills to two-thirds of its capacity or drains to one-third of its capacity with the "RC time constant" equation which (discussed earlier in this chapter in the section "Turning Things On and Off").

The RC time constant for filling the capacitor is

$$T1 = (R1 + R2) \times C$$

The RC time constant for draining the capacitor is

$$T2 = (R1) \times C$$

In this circuit, R1 and R2 determine how fast the capacitor charges and discharges. The extent to which the capacitor is filled determines the voltage on the trigger and threshold pins. When the voltage reaches two-thirds of V+, the IC is triggered causing the output to change from +V to 0 (zero) volts and the charge on the capacitor to drain through the discharge pin. As the capacitor drains, the voltage to the IC trigger and threshold pins drops. When the voltage gets to one-third of +V, the IC is triggered again to bring the output voltage from 0 (zero) to +V and to allow the capacitor to charge back up to two-thirds of V+, at which point the cycle starts again.

This sequence repeats and produces the digital waveform in Figure 7-14. Changing the values of R1 and R2 changes the RC time constants, and hence the shape of the digital waveform changes. In Chapter 14, you can see how to use this type of output in actual projects.

Part IV
Getting Your Hands Dirty

In this part . . .

In this part you roll up your shirt sleeves and get your nails dirty (electronics types love this stuff). You explore the joy of soldering a circuit together so that all the parts don't fall out when you pick it up. Soldering isn't hard, but it takes some practice and skill, not to mention the right tools and supplies. We tell you everything you need to know about how to be a solder pro.

You also learn how to use a multimeter to test circuits and figure out what the heck is wrong with the project you just made. A multimeter lets you get into the mind of a circuit, and see how it ticks (or doesn't tick, as the case may be). And we tell you the basics of using two nifty, but optional, test gadgets: the logic probe and the oscilloscope.

Chapter 8

Everything You Need to Know about Soldering

*S*oldering is the method you use in your electronics projects to assemble components on a circuit board to build a permanent electrical circuit. Instead of using glue to hold things together, you use small globs of molten metal called *solder*. The metal not only provides a physical joint between the wires and components of your circuit, it also supplies the circuit with the conductivity it needs to work.

Despite working with temperatures in excess of 700 degrees Fahrenheit, soldering is fun and generally safe (if you observe the normal precautions). You need only a minimum of tools and supplies, most of which you can purchase locally at hardware and home-improvement stores.

To Solder or Not to Solder

Before getting into the how's of soldering, we'll talk about the why's. You don't need to solder all electrical circuits. You can use the solderless breadboard method instead (meaning you insert components and wires into the holes on a breadboard but don't permanently affix them) to construct a working circuit, especially if you're just experimenting.

You're better off using a solderless breadboard if

- **You're just playing around with ideas.** You may want to pull components out and try new ones or experiment with different ways to connect things. Although you can solder your circuit designs, any mistakes or changes require unsoldering, which involves melting the solder and removing it with a little suction device. With the solder removed, you can then pull out the component.

- **You're testing your circuit to be sure that it works properly.** Even the best electronics experts try out their ideas before committing them to permanent soldered status. With a solderless breadboard, you can more readily make changes to improve the circuit.

- **You don't need or want a permanent circuit.** You can build a simple flasher circuit for a miniature Christmas tree display on a solderless breadboard. The flasher circuit is a temporary circuit used for just a few weeks out of the year. Come January 1, just tear it apart and reuse the components (and breadboard) for something else . . . like a Valentine's Day blinking heart!

- **You want to customize the circuit as you work with it.** Rather than building several options into one circuit, you may want to reconfigure it to change its behavior. You can often make this change by switching out basic components, such as resistors and capacitors. You can make such changes in seconds with a solderless breadboard.

On the other hand, any circuit that requires permanence or that may be damaged by ordinary handling almost always needs soldering. Here are some specific examples of when to solder the circuit, using some type of circuit board:

- **Solder the circuit if handling, motion, or vibrations may work the connections loose.** This situation may be the case, for example, for any circuit that you mount in a car, such as a wind speed indicator, or the electronic eyes you mount on a robot.

- **A properly-soldered circuit lasts much longer than one mounted on a solderless breadboard.** If you plan on using the circuit for more than a few weeks, permanently solder it.

- **Soldered circuits are less prone to the effects of stray capacitance.** Long lead lengths on the components and in the construction of the solderless breadboard itself, cause stray capacitance (where electric fields occur between the two leads and energy is stored, as it is in a capacitor). Stray capacitance can affect the operation of circuits in unpredictable ways. You notice these effects most with circuits that already rely on capacitors for signal timing.

- **We highly recommend soldering for any circuit, such as a power supply that directly plugs into a wall socket.** You have less risk of shock or fire if you securely solder all wiring and components to a rigid circuit board or terminal.

✔ **Soldering is also the best method for any circuit that uses high currents, such as a motor driver for your remote control race car.** Solderless breadboards can't support more than an amp or two of current; if you use more current, things start melting.

Things You Absolutely, Positively Need for Soldering

You'll be glad to hear that you only need some pretty simple tools for soldering. You can purchase a basic, no-frills soldering setup for under $10, but the better soldering tools cost you more.

Here's a rundown of the basic soldering tools that you need:

✔ **Soldering pencil.** A *soldering pencil,* also called a *soldering iron,* is a wand-like tool that consists of an insulating handle, a heating element, and a polished metal tip (see Figure 8-1). It's pretty obvious why you call it a pencil. It looks a lot like a good old #2. The more generic *soldering iron* takes many forms, including a large gun-like appliance that was common from the 1940s through the '60s. Don't use these big soldering irons with modern electronics because they produce way too much heat. For standard electronics work, you want a soldering pencil rated at 25 to 35 watts. A 27- or 30-watt pencil is ideal. Be sure to get a soldering pencil with a replaceable tip. That way when the old tip gets all worn out, you can easily replace it.

✔ **Soldering pencil stand.** The better soldering pencils come with a stand, but many low-cost ones don't. You want to get your hands on one if your soldering pencil is stand-less. A stand holds the hot soldering pencil when you're not using it, and it helps prevent accidents. Trust us: You really don't want the hot soldering pencil to roll off the desk and onto your lap. *Ouch!*

✔ **Solder.** *Solder* is the soft metal that the heat of the soldering pencil melts. The ideal solder for working with electronics is called *60/40 rosin core.* This name refers to the fact that the solder contains 60 percent tin and 40 percent lead (the exact ratio can vary a few percentage points) and has a core of rosin flux. This *flux,* which is a wax-like substance, helps the molten solder flow around the components and wire, and it assures a good joint. Solder comes in various diameters. A 0.062-inch diameter is common. You may have more trouble using solder thicker than about 0.080 inches on small circuits.

Sometimes you see solder sold by gauge, and sometimes by an actual diameter, expressed in inches (or millimeters, for countries using the metric system).

Figure 8-1:
Some
soldering
pencil
models are
temperature-
adjustable,
and come
with their
own stand.

When you see the wire as a gauge, it's important to note that the smaller the number, the larger the solder diameter. Wander over to Chapter 5 for more about wire gauge.

Here are two common solder gauges, along with their actual dimensions.

Common Solder Gauges

0.031"	22 gauge
0.062"	16 gauge

Soldering releases toxic fumes. You can buy lead-free solder to avoid the effects of lead poisoning. These solders contain other mixtures of metal, such as 95 percent tin and 5 percent antimony. But almost any soft metal that you can find in solder — such as lead, bismuth, indium, or antimony — is toxic to one degree or another. Always solder in a well-ventilated area, regardless of the composition of the solder! As of this writing, no one has come up with a completely non-toxic solder. *Don't use silver solder or any other solder not specifically intended for electronics, especially solder designed for copper plumbing pipes.* These solders may not provide the same conductivity as standard 60/40 rosin core, and they may cause corrosion or leave contaminants that could make the circuit completely inoperable.

Additional items useful for soldering that you can get in most hardware stores include

✔ **Wetted sponge:** For wiping off excess solder and flux from the hot tip of the soldering pencil. Just a basic (but clean!) kitchen sponge does the trick. In a pinch, you can fold up a paper napkin, dampen it, squeeze out any excess water, and use it like a sponge.

✔ **4X to 6X magnifying glass:** For inspecting your work. After soldering, always check your solder joints to make sure they're clean and well-formed and that no solder touches adjacent wires or circuit board pads.

✔ **Solder sucker:** For removing excess solder. The sucker is a spring-loaded vacuum. To use it, melt the solder that you want to remove and then quickly position the sucker over the molten glob. Activate the sucker, and it removes the extra solder.

✔ **Rosin flux remover:** Available in a bottle or spray can, use this after soldering to clean any remaining flux to prevent it from oxidizing your circuit.

✔ **"Third hand" clamp:** Soldering would be a lot easier if everyone had three hands. Alas, most people are born with only two, so the next best thing is a small, weighted clamp that holds the parts while you solder. You may hear these clamps referred to as "helping hands" or simply a "third hand." Figure 8-2 shows one of these clamps. You can purchase them with or without an integrated magnifying glass.

Figure 8-2:
A so-called "third hand" clamp helps you hold components while soldering. Get the kind with a built-in magnifier.

Choosing just the right soldering pencil

The basic soldering pencil you use for electronics work (shown in Figure 8-3) is composed of a removable tip, and a 25- to 35-watt heating element. The basic soldering pencil gets you soldering your circuits together, but not in style. Although it costs a little more, a soldering pencil with an adjustable temperature control gives you a better result. With these controls, you dial in the best temperature for the job.

If your soldering pencil doesn't come with one, you need to get a separate stand for it. Soldering stands are inexpensive, so don't cheap out and just lay the soldering pencil on your desk while you're working. You're bound to burn your project, your desk, or yourself!

Figure 8-3:
The basic soldering pencil, in all its glory.

Although some of the higher-end variable-temperature soldering pencils come with a digital readout, showing you the actual temperature at the tip, you don't really need this feature for basic electronics work, though it's nice to have as you build bigger and more complex circuits. With some practice and experience, by watching how quickly the pencil melts the solder you can figure out how to gauge the proper temperature setting.

Select a soldering pencil that comes with a grounded cord and plug. Many people consider a grounded electrical plug to be safer, in the event that the soldering pencil comes into contact with a live electrical circuit.

Selecting a soldering tip

The soldering tip attaches (it usually screws on) to the end of the heating element. The tip does the actual soldering. You can choose from literally hundreds of soldering tips, but don't let that confuse you. For most electronics work, you want a small conical or chiseled tip. These kinds of tips come in various sizes: ⁶⁄₆₄-inch through ¹⁶⁄₆₄-inch tips do the job for most electronics work.

You often can't swap soldering tips among different brands of soldering pencil or even different models by the same manufacturer. Be sure to purchase the correct tip for your make and model of soldering pencil.

Replace soldering tips as they show signs of wear. Look for corrosion, pitting, or plating that is peeling off. Replace tips that no longer provide adequate heat or else your solder joints won't be as strong as they should be because, when soldering tips get old, they don't pass as much heat. That can also slow down your work. Eventually, if the tip gets really worn, it won't ever make enough heat to melt the solder.

Preparing Your Soldering Pencil

Before soldering, make sure you have all your tools within easy reach and then follow these steps:

1. **Dampen a small sponge or a folded-up paper towel. Squeeze out any excess water.**

   You want it to be damp, not soaked.

2. **Place the soldering pencil securely in its holder and plug it in.**

3. **If you have the adjustable type of pencil, turn the heat to approximately 675 to 750 degrees Fahrenheit.**

4. **Wait for the tool to reach proper temperature — usually within 60 seconds for most 25- to 30-watt soldering pencils.**

   Many soldering pencils with a temperature sensor let you know when they reach the proper temperature by lighting or blinking an indicator.

 If the tip is new, *tin* it before soldering. Tinning is recommended because it helps prevent solder from sticking to the tip and forming into an ugly globule. If the globule comes off onto your circuit, then you could get a short. You tin the tip by heating up the pencil to full temperature and applying a small amount of solder to the tip. Wipe off any excess solder with a moistened sponge or towel. Periodically use this same technique to keep the tip clean. You can also purchase soldering tip cleaners if dirt becomes caked on and you just can't get it off during regular tip re-tinning.

Successful Soldering

Successful soldering requires that you follow some simple rules and get a lot of practice. Keep the following in mind as you solder:

✔ **The cleaner the metal surface, the better the solder sticks to it.** Clean etched circuit boards and wire ends with isopropyl alcohol. Let surfaces dry thoroughly before soldering: You don't want them to catch on fire!

✔ **Hold the soldering pencil at a 30- to 45-degree angle to the work surface.** (See Figure 8-4.) If you're using a chiseled tip, the flat of the chisel should rest firmly against the surface of the joint that you're soldering.

✔ **Always apply the heat of the tip to the item that you're working on, not to the solder.** If you're soldering a wire into the hole of a circuit board, for example, touch the tip to both the wire and the pad, like the tip in Figure 8-5. Wait a few seconds and then apply solder to the heated area. Immediately remove the heated tip after the solder flows.

✔ **Apply just the right amount of solder.** Too little, and you form a weak connection; too much, and the solder may form globs that can cause short circuits.

✔ **You know you have just the right amount of solder when it forms a raised area (called a *fillet*) between the wire and the circuit board.**

✔ **Avoid applying more solder to an already-soldered joint**. This added solder can cause what's known as a cold solder joint and the result could be that your circuit simply won't work. (Check out the following section, "Avoiding Cold Solder Joints like the Plague," for more on this soldering no-no.)

Figure 8-4:
Hold the soldering pencil at a 30- to 45-degree angle.

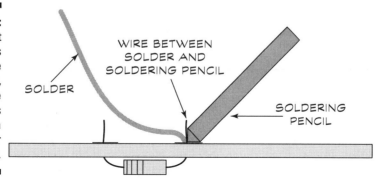

Figure 8-5:
Apply heat
to the parts
you're
soldering,
not to the
solder! This
leads to a
better
solder joint.

WIRE BETWEEN
SOLDER AND
SOLDERING PENCIL

SOLDER

SOLDERING
PENCIL

You can damage many electronic components if you expose them to prolonged or excessive heat. Apply the soldering pencil only long enough to heat the work for proper soldering — no more, no less.

When soldering electronic components that are very heat sensitive, use a *clip-on heat sink*. These sinks look like miniature aluminum pliers, with a spring-loaded clamp that you attach securely to the component you want to protect. The sink draws off heat and helps prevent the heat from destroying the component. Clip the sink to the wire that you're soldering, as near to the component itself as you can. Of course, you still have to exercise caution, even when using a heat sink.

Avoiding Cold Solder Joints like the Plague

A cold solder joint happens when solder doesn't properly flow around the metal parts. Cold joints are physically weaker than properly made joints, and they don't conduct electricity as well. You can often (but not always) identify cold joints just by looking at them. A cold joint typically has a dull appearance rather than the shiny, uniform look of a normal joint. And the solder may form jagged peaks rather than having an all-around smooth surface.

Many things can cause a cold solder joint, such as:

✔ **You move the work as the solder is cooling.** Avoid all movement until the solder cools beyond the plastic phase. (The *plastic phase* occurs when the solder is still partially liquid and not yet hardened.) If you accidentally jiggle the wire or component, quickly re-apply the tip of the soldering pencil to reheat the solder back to its liquid state.

✔ **The joint is dirty or oily.** Be sure to keep all metal-to-metal contacts clean.

✔ **You don't heat the work to the proper temperature.** Be sure that you have the work hot enough to melt the solder to a somewhat runny liquid.

✔ **You apply the solder to the soldering pencil and not to the heated work.**

✔ **You solder and then resolder the work.** During soldering, the original solder isn't heated enough. It's best to remove as much of the old solder as possible and then completely remake the joint with all new solder.

When you experience all but the first item in the list above, you need to unsolder the old joint and re-apply fresh solder. Avoid simply reheating the solder; you rarely get the proper flow, and you're likely to end up with another cold joint. See the section "Unsoldering and Resoldering," later in this chapter for the proper procedure.

Avoiding Static Discharge While Soldering

The soldering process can generate electrostatic discharge (ESD), which can cause damage to sensitive electronics components and ruin your whole day. Simply handling the components and circuit board can lead to static, as can the soldering pencil itself. You can't totally eliminate static discharge, but you can minimize it.

Not all electrical components are static-discharge sensitive. But, for safety's sake, you should develop static-safe work habits when handling any electrical components. For a list of major electronic components and their level of susceptibility to damage from static discharge, go to Chapter 2 and take a look at Table 2-1.

Thwarting discharge before it begins

Here are a few things you can control to reduce the danger of static discharge:

✔ What you wear can have a great impact on the amount of static that develops around you. Synthetic clothing tends to generate static. Instead, wear natural cotton.

✔ If you're working indoors on carpet, wear shoes instead of going barefoot or in socks.

✔ Wear an anti-static wrist band whenever possible. A wire from the wrist band attaches to any grounded object and helps to draw off static from your body. See Figure 8-6 for a picture of an anti-static wrist band.

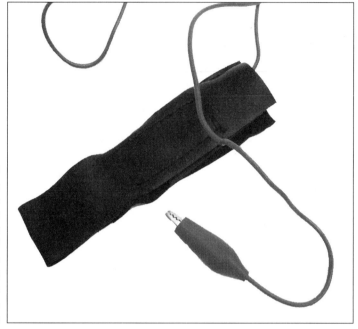

Figure 8-6:
An anti-static wrist band helps draw off damage-causing static from your body.

Stocking up on anti-static supplies

You can go anti-static all the way by keeping these items handy:

- ✔ **Anti-static work mat:** Place an anti-static work mat on the floor under your worktable, especially in carpeted rooms. This simple addition prevents static buildup as you shuffle your feet.

  Use an anti-static mat on your worktable, too. Avoid the nylon carpet that electronics newbies typically use as a work mat; the carpet provides a nice cushion for your projects, but it can generate static and even melt.

- ✔ **Anti-static spray:** If you can't find an anti-static work mat, get yourself a bottle of anti-static spray. Some sprays actually attract dirt, so you may need to clean your carpets more often.

- ✔ **Anti-static bags:** Keep static-sensitive electronic components in anti-static bags until you're ready to use them. (Most components come in these bags, or they're stuck into foam that has anti-static properties.) Minimize handling components whenever possible.

Static buildup can turn into a serious problem in dry weather. If you live in a dry climate, you need to take extra precautions against ESD. You can buy humidifiers, electrically grounded anti-static mats, and other ESD control products to reduce static, but these items cost serious money. Still, it's cheaper than moving to the rainforest.

All carpets aren't created equal

Some types of carpets are more prone to static than others. Regular nylon carpet can generate massive amounts of static as you walk over it. Be sure to drain the static from your body by touching a doorknob or the metal portion of a grounded appliance before touching any of your electronic components or tools. Designers make the fiber in many types of commercial carpeting low-static. Consider this option if you need carpeting for your workroom. As an alternative, you can buy small remnant pieces of low-static or anti-static commercial carpeting from a carpet dealer and place a remnant in front of your workbench.

Unsoldering and Resoldering

Even the experts sometimes insert a component backwards! It's inevitable that you occasionally need to undo a solder joint to fix mistakes or to clean up a cold solder joint. When this situation happens, you need to remove the solder at the joint and apply new solder.

You can use a desolder pump, solder wick, or both to remove solder from the joint.

Use solder wick (also called solder braid) to remove hard-to-reach solder. The *solder wick* is really a flat braid of copper. It works because the copper absorbs solder more easily than the tin plating of most components and printed circuit boards. Exercise care when using solder wick because if you touch the hot braid, you can get a serious burn.

I prefer the *desolder pump*. The desolder pump works by sucking up the excess solder with a vacuum. Desolder pumps come in two basic styles: spring-loaded plunger and bulb. They both work the same way in that they suck up molten solder, but the spring-loaded plunger is a little easier to use. That's because the bulb requires a little more manual dexterity, as you need to squeeze it one or more times with one hand, while holding the soldering pencil over the joint you're melting with your other hand

Putting a spring-loaded plunger desolder pump to work

Here are the steps you should follow to use a spring-loaded plunger desolder pump:

1. **Depress the plunger and then position the nozzle over the joint that you want to remove.**

   See Figure 8-7 for an example.

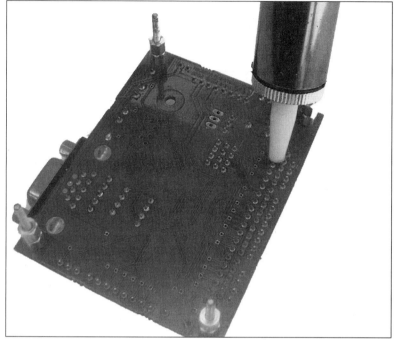

Figure 8-7:
Place the nozzle of the desolder pump close to or even touching the solder that you want to remove.

2. **Carefully position the soldering tip into the joint to heat the solder.**

   Be careful not to touch the end of the desolder pump or you may damage the nozzle.

3. **When the solder begins to flow quickly, release the plunger to suck up the solder.**

4. **Depress the plunger one more time to expel the solder from the pump into a receptacle.**

   It's a good idea to do this over a wastepaper basket so that you don't leave bits of solder debris on your workbench or in your project!

5. **Repeat Steps 1 through 4 as needed until you remove as much of the old solder as possible.**

This bulb desolder pump definitely sucks

Bulb desolder pumps work a lot like the spring-loaded variety, except that you squeeze the bulb to suck up the solder. You may have some problems using these pumps unless you mount the bulb on the soldering pencil. You can find some soldering pencils especially designed for desoldering that have this arrangement.

Be sure to release the plunger when you're done using the desolder pump. The tool lasts longer that way, and it holds its vacuum better. Don't store the desolder pump with a cocked plunger because the rubber seal can become deformed. If the seal gets deformed, the pump doesn't create enough vacuum to suck up any solder.

When you've removed the old solder, you may reapply solder to the joint, following the instructions that you can find in the section "Successful Soldering," earlier in this chapter.

Soldering Tips and Techniques

Soldering isn't rocket science. Still, you'd be wise to consider these tips, techniques, reminders, and suggestions:

- ✔ Remember, cleanliness is king. Be sure that you keep all surfaces that you're going to solder free of dirt and oils. Otherwise, you may end up with a weak soldered joint, or one that impairs conductivity.

- ✔ Metal dental picks make for good soldering tools. You can use the picks to clean the work area prior to soldering and to scrape away excess solder from a joint. You can get used (but clean!) dental tools from a variety of mail-order surplus stores, including American Science & Surplus. (See the appendix for more details.)

- ✔ Store your spool of solder in a resealable plastic bag. Doing this little chore helps keep the solder clean. It may pick up dirt and oils if you simply throw it into your toolbox. If the spool does get soiled, clean it with isopropyl alcohol before using it.

- ✔ Allow the soldering pencil to cool completely before putting it away. If you don't use the soldering pencil often, put the cool pencil in a large plastic bag to keep it clean.

- ✔ If you've grounded the electrical cord of your soldering pencil, be sure to plug it into a grounded outlet. Don't cut off the ground connector or bypass the grounding by using an adapter. The manufacturers include the ground for safety.

- ✔ After soldering, and when you're sure that your circuit operates properly, spray or brush on something called flux cleaner. This chemical removes the left-over rosin residue, also called flux.

- ✔ You use the same general techniques we've described here to solder surface mount components (teeny-tiny components that don't have wire leads). With practice, a steady hand, and a good eye (or a good magnifying glass!), you can solder many types of surface mount components. Don't try this right away if you're new to soldering, though. Get some experience under your belt first.

Chapter 9

Making Friends with Your Multimeter

A multimeter is to an electronics geek as an oxygen tank is to a scuba diver. Sure, you can hold your breath underwater, but not for long; you soon have to come up for air. As a builder of cool electronics gadgets, you can only experiment for so long before you need a multimeter to take you the rest of the way.

With this one handy tool, you can not only verify proper voltages but also test whether you have a short circuit or if there's a break in a wire or connection. You'd be surprised how much troubleshooting you can do with just these simple tests, and a multimeter does them all.

In this chapter, you learn the basics of using a multimeter to perform important checks on electronic circuits and parts. These tests help you determine if everything is A-OK or if you have a problem that Houston should know about.

The Basics of Multimeters

The multimeter, also called a volt-ohm meter (or VOM), is the basic tool for anyone working in electronics. You can see a fairly typical modern multimeter in Figure 9-1.

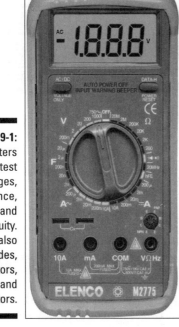

Figure 9-1:
Multimeters
test
voltages,
resistance,
current, and
continuity.
Some also
test diodes,
capacitors,
and
transistors.

You use a multimeter to take a variety of electrical measurements — hence
the term "multi." With this one tool, you can

- Measure AC voltages
- Measure DC voltages
- Measure resistance
- Measure current going through a circuit
- Measure continuity (whether a circuit is broken or not)

And, depending on the model, you may also be able to test the operation of
diodes, capacitors, and transistors to see if they're good.

All multimeters come with a pair of test leads, one black and one red (black is
for the ground connection; red is for the positive connection). Each test lead
comes equipped with a metal probe. For small, pocket units the test leads
come permanently attached to the meter. On larger models, you can unplug
the test leads.

If you don't already own a multimeter, you should seriously consider buying
one. It's well worth the relatively low cost, considering how much you'll use
the meter. Prices for new multimeters range from $10 to over $100. The main

difference between the high- and low-priced meters is the features that you get, such as built-in testing features for capacitors and transistors. Shop around and compare features and prices. Know that, whatever you buy, you're going to have it for years to come. Consider getting the best multimeter that you can afford so that, as your projects grow more complex, your multimeter can keep up.

Remember: Safety First!

Most tests using a multimeter involve low voltage and resistance, both of which can't hurt you much. But sometimes you may need to test high voltages, such as the input to an AC-operated power supply. In a case such as this, careless use of the multimeter can cause serious bodily harm. Even when you're not actively testing a high voltage circuit, dangerous current may be exposed if you work on certain electronics equipment, such as a radio, hi-fi, or VCR.

Remember this: If you ever need to work with an electronics project that uses house current (117 volts in the US; 220 volts in many other countries) and you touch a live AC wire, you can seriously hurt or even kill yourself. Always exercise caution when handling electronic equipment and electric wires. Be especially careful to keep your fingers away from the metal tips of the meter test leads. The test leads are the wire probes that you use to connect the multimeter to your circuit. If you handle the probes carelessly during testing you may get a serious shock.

Never blindly poke around the inside of a circuit with the leads of a multimeter in an attempt to get a reading. Apply the test leads only to those portions of the circuit that you are familiar with. One safe method for using a meter is to attach a clip on the black (negative or common) lead and connect that lead to the chassis or circuit ground. Use one hand to apply the red (positive) lead to the various test points and stick the other hand safely in your pocket. With one hand out of commission, you're less likely to receive a nasty shock, even if you aren't watching what you're doing.

Which to choose: Digital or analog?

Multimeters come in two general flavors: digital and analog. These names don't mean that you use one on digital circuits and the other on analog circuits. It's a bit simpler than that:

- Digital multimeters use a numeric display, like a digital clock or watch.

- Analog multimeters use the old-fashioned — but still useful — mechanical movement that uses a needle to point to a set of graduated scales. Figure 9-2 shows an example of an analog multimeter.

Figure 9-2:
An analog multimeter uses a needle to indicate voltage, current, and other values.

Digital multimeters used to cost more than their analog cousins, but the price difference has evened out. Digital meters are fast becoming the standard. In fact, although some manufacturers still make them, you have a hard time finding a good analog meter anymore.

If you really, really want an analog multimeter, you may as well get a top-notch one. But top-of-the-line analog multimeters can cost you a pretty penny if you purchase them new. An alternative is to buy one through eBay. Try a search for *Simpson meter 260* — the Simpson Model 260 was one of the most popular meters ever produced. They may look like relics by today's standards, but as long as no one has abused the meter, it should do all the basic tasks that you need.

Traditionally, users have a harder time with analog multimeters because you have to select the type of testing (voltage, current, or resistance), as well as the range. You must then correlate the results using the proper scale on the meter face and estimate the reading as the needle swings into action. In contrast, digital multimeters display the result as a precise number. Those numbers help take away the guesswork.

Taking a Close-Up Look at Multimeters

Multimeters aren't particularly complicated, but the following sections give you some factoids that you should know before you choose one or use one. We cover the basic functions shared by all meters, some of the dials that provide meter readouts, issues related to meter accuracy and the supplies that come with the meter. You also need to know whether the meter automatically adjusts itself to display the most accurate result possible (called *auto-ranging*) and whether it has special testing features for checking diodes, capacitors, and transistors.

Basic features of every meter

Stripped down to its skivvies, a multimeter's purpose is to take the three basic measurements of electronics: voltage, current, and resistance.

Hello, any voltage or current in there?

You test voltage and current with a circuit powered up. Typical voltage and current tests include

- ✔ **Checking the voltage level of a battery.** You can even check the voltage when you're using the battery. In fact, many consider this test more accurate when the battery is providing power — what electronics folks call under load.

- ✔ **Determining if a circuit or component is drawing too much current.** If the circuit has more current going through it than it's designed to handle, then the components may get overheated and you can permanently damage your circuit.

- ✔ **Verifying that the proper voltage reaches a component, such as a light-emitting diode or switch.** These kinds of checks can help you pinpoint the location of a problem in your circuit. You use multimeter tests to narrow down the field of suspects until you find the culprit causing all your headaches.

Checking out the resistance movement

You almost always test resistance (measured in ohms, as we talk about in Chapter 1) with the circuit unpowered. Resistance tests may involve an entire circuit or just an individual component. You can check up on wires, resistors, motors, and many other kinds of electronic doodads.

Beep beep goes the continuity test

A feature found on many multimeters, like the one shown here, is audible continuity testing. To use this feature, you turn the meter's control dial (more about dial turning in the section "Making sense of all the inputs and dials") to Continuity or Tone

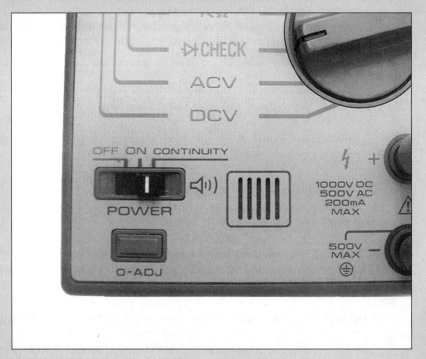

You may find this feature handy when you check the wiring of a circuit. If a wire or connection has continuity (a shorted circuit), the meter beeps. If the wire or connection doesn't have continuity (an open circuit), the meter stays silent. The audible tone gives you a handy way to check a whole circuit without having to keep your eye on the multimeter. Most meters made these days have this feature, and we recommend it.

Resistance, or the absence of it, can reveal short circuits and open circuits; so-called *continuity* of electrical components. When you perform these tests, a shorted circuit shows zero (or virtually zero) resistance and an open circuit shows infinite resistance. You can use continuity tests to check for breaks in wires.

Here are some other tests you can perform with a multimeter that rely on resistance:

- ✔ **Fuses:** A blown fuse shows an open circuit.

- ✔ **Switches:** Flipping the switch should alternate the multimeter's reading between zero (shorted) and infinite (open) resistance.

- ✔ **Circuit board traces:** A bad copper trace on a printed circuit board acts like a broken wire and shows up as infinite ohms (open circuit) on the multimeter.

- ✔ **Solder joints:** A bad joint may read as an open circuit on the multimeter, showing infinite resistance.

Making sense of all the inputs and dials

Check out Figure 9-3 to see the main points of interest on the typical multimeter. Here's what they all mean:

- ✔ **Meter face or digital readout:** Analog multimeters have a meter face consisting of a set of graduated scales and a precision needle indicator. A digital multimeter has a numeric readout.

- ✔ **Function knob:** Dial the knob to the test that you want to perform: Voltage, Current, Resistance, or whatever. On meters without an auto-ranging feature, you also typically use the function knob to set the maximum range of the value that you want to test. If you set the maximum range to be just higher than the value you are testing — whether voltage, resistance, current, or whatever — you are assured of the most accurate reading possible. If your meter does have an auto-ranging feature, it will automatically adjust itself to give you the most accurate reading.

- ✔ **Test lead inputs:** At a minimum, the multimeter has a + (positive) and − (negative or common) lead input. You insert the test leads into these inputs. Some meters have additional inputs for high current testing (usually marked A, for amperage) and special sockets for testing transistors and capacitors, as you can see in Figure 9-4. *Note:* Many small, pocket multimeters have the leads permanently attached.

- ✔ **Zero-set control:** On analog meters without an automatic zero feature, designers provide a rotating knob so you can adjust the needle to 0 (zero) ohms before use. Some digital meters have a button that, when you press it, sets the meter to zero.

Figure 9-3:
Your
multimeter
may not look
like this one,
but odds are
yours has
similar
features.

Figure 9-4:
Most
meters
provide
additional
input
sockets for
testing
capacitors
and
transistors.

Accuracy, resolution, and sensitivity

The accuracy of a multimeter represents the maximum amount of error that occurs when it takes a measurement. For example, the multimeter may be accurate to 2,000 volts, ±0.8 percent. A 0.8-percent error with the types of voltages used in DC-operated circuits — typically 5 to 12 volts DC — measures only about 0.096 volts. For hobby electronics projects, you don't need a more precise level of accuracy. As you compare the accuracy of multimeters, bear in mind that just about every model of meter gives the hobbyist the results that he or she needs.

Digital meters have another type of rating, this one more commonly called resolution. The number of digits in the display determines the smallest change that the meter can register. Most digital meters designed for hobbyists have 3½ digits, so they can display a value as small as 0.001 (the half digit appears as a 1 on the far left of the display). The hobbyist's meter can't accurately represent anything less than 0.001. For most hobby-level electronics projects, you don't need to worry about this.

Resolution in digital multimeters is also a function of analog-to-digital converter (ADC) electronics. An *ADC* converts an analog signal to a digital one. Many consumer-grade multimeters use a 12-bit ADC. Without getting into all the technical mumbo-jumbo, a 12-bit ADC can take any analog signal and convert it into 4,096 discrete steps. (These discrete steps are necessary because of the way digital circuits work. In the digital world there can't be any in-between or "sort of" values.) Meter manufacturers select an ADC with a resolution that works with the number of display digits on the device. A 3½ digit digital readout displays the values of a 12-bit ADC just about right.

Along the lines of accuracy and resolution, you need to consider the *specification for sensitivity.* This phrase means the smallest value that a meter can meaningfully detect when you use it under normal conditions.

- Quality digital multimeters sport a maximum sensitivity of about 1 microvolt (AC or DC); that's one millionth of a volt. The lower the value, the better the sensitivity.

- Quality analog multimeters offer a maximum sensitivity of about 20,000 ohms per volt, typically shown as 20KΩ/V. The higher the ohms value, the better the sensitivity.

The well-stocked multimeter

The typical multimeter doesn't come with a lot of accessories, but you need to have a few. We cover the necessities in the following sections.

Manual

The manual for your multimeter may be just a single sheet of paper with little more than a picture, or you may get a small booklet with step-by-step instructions. Either way, be sure to at least browse through the manual. It contains important safety precautions, as well as a run-down of features and specifications for that meter model.

Test leads

The test leads included with most inexpensive multimeters aren't of the highest quality, so you may want to purchase a better set. You may want to get the type with coiled leads because they stretch out to several feet, yet recoil to a manageable length when not in use. Figure 9-5 shows some examples of coiled leads.

Standard leads with their pointed metal probes work fine for most routine testing, but some measurements may require the use of a clip lead. These leads have a spring-loaded clip on the end; you can clip the lead in place so that your hands are free to do other things. The clips are insulated to prevent touching the metal against another part of the circuit.

If your multimeter doesn't come with clip leads, you can buy some clip-on attachments that fit over regular test leads.

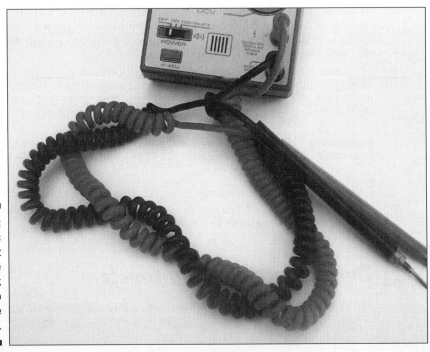

Figure 9-5:
Coiled leads stretch out during use but shrink back to normal size for storage.

Extra fuse

Most multimeters use an internal fuse to protect themselves against excessive voltage or current. The better multimeters come with a spare fuse. If yours doesn't have a spare, purchase one when you buy the meter. That way, you have it available when you need it.

Some meter fuses are specially made, and replacements can cost you a bundle. You may want to check the price of replacement fuses before you purchase the meter!

Batteries

Except for really old analog models that only test voltage or current, all multi-meters come equipped with a battery of one type or another. The most convenient multimeters use a standard-size battery, such as a 9-volt or AA cell. Pocket meters typically use a coin-type battery. If your local supermarket or drugstore doesn't carry replacement batteries for your meter, try Radio Shack or a photographic supply store.

The batteries in multimeters tend to last a long, long time — that is, unless you forget to turn the meter off after using it. The batteries in multimeters can often last a year or longer under typical use. But eventually the battery dies, so be sure to keep a spare battery handy. We prefer alkaline batteries over standard-duty zinc cells, as they last longer.

If your multimeter uses a specialty battery, consider storing the spare in its original packaging in your refrigerator. It lasts longer that way. Take it out of the refrigerator a day before you plan to use it. That allows the battery to slowly come up to room temperature.

Nickel-cadmium and nickel metal hydride rechargeable batteries put out a slightly lower voltage than alkaline batteries of the same size. Most multi-meters don't have a problem with this lower voltage. However, some meters may stop working or may give erratic or erroneous results when powered by a rechargeable battery. Check the manual that comes with your multimeter to be sure that your meter can handle rechargeable batteries.

Maximum range: Just how much is enough?

There's a limit to what a multimeter can test. You call that limit its *maximum range*. These days, most consumer multimeters have more-or-less the same maximum range for voltage, current, and resistance. Any meter that has the following maximum ratings (or better) should work just fine for your hobby electronics:

DC volts: 1000 V

AC volts: 500 V

DC current: 200 mA (milliamperes)

Resistance: 2 MΩ (two megohms, or 2 million ohms)

Home on the automatic range

Most analog multimeters, and many digital ones, require that you select the range (see Figure 9-6) before the meter can make an accurate measurement. For example, if you're measuring the voltage of a 9-volt transistor battery, you set the range to the setting closest to, and above, 9 volts. For most meters, this means you select the 20 or 50 volt range. You then read the voltage on the meter.

Be sure to read the result from the proper meter scale. If you select the 20-volt range, for example, you must use the 20-volt scale. Otherwise, you end up with inaccurate results.

You shouldn't find manually setting the range of your meter complicated, and the extra effort can't kill you. But these days, automatic ranging, especially for digital multimeters, is all the rage. So-called auto ranging meters don't require you to first set the test range. This feature makes them inherently easier to use and a little less prone to error. When you want to measure voltage, you set the meter function to Volts (either AC or DC) and take the measurement. The meter displays the results in the readout panel. Meters with an automatic ranging feature, like the one in Figure 9-7, don't require a separate range knob.

What if you need to test higher currents?

Most digital multimeters can measure current only less than one amp. The typical digital multimeter has a maximum range of 200 milliamperes. Attempting to measure substantially higher currents may cause the fuse in the meter to blow. Many analog meters, especially older models, support current readings of 5 or 10 amps, maximum.

You may find analog meters with a high ampere input handy if you're testing motors and circuits that draw a lot of current. If you have only a digital meter with a limited milliampere input, you can still measure higher currents indirectly by using a low-resistance, high wattage resistor. You can read more about this kind of resistor in Chapter 7.

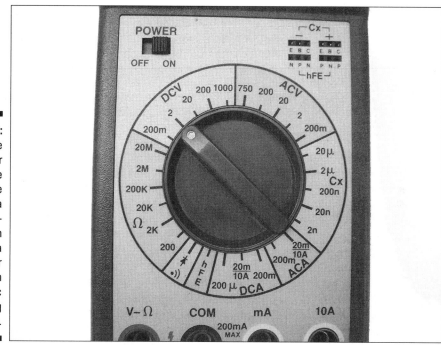

Figure 9-6:
Dial in the proper range before taking a measurement when using a multimeter without an automatic ranging feature.

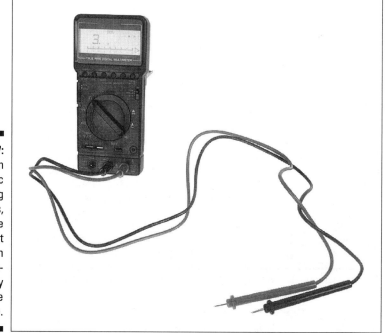

Figure 9-7:
On automatic ranging multimeters, setting the desired test function automatically selects the range.

Whether analog or digital, the meter indicates an over range if the voltage or other measurement is too high for the meter to display. A digital multimeter typically shows over range as a flashing 1 (or OL). An analog meter shows over range as the needle going off the scale. If the meter is auto-ranging, and you see the over range indicator, it means that the value is too high to be measured by the meter. Such an over range indication is common when testing continuity. It simply means the resistance is so high that that meter cannot register it, even at its highest range setting.

When using an analog multimeter, avoid over range conditions because these conditions can damage the precision needle movement. For this reason, always dial in the highest scale that you believe you need when using an analog meter and then work your way down. This approach avoids the needle slamming against its stops (the upper limit reading) in an over range.

Extra nice-to-have functions

As we discuss in the section "The Basics of Multimeters," earlier in this chapter, all standard multimeters let you measure AC volts, DC volts, current, and resistance. Beyond these functions, digital multimeters vary in the number and type of functions that they provide. Here are some extra functions that can make the testing process a little easier and a bit more accurate:

✔ **Test the operation and value of capacitors.** Because test leads can influence capacitance readings, most multimeters with a capacitor-testing feature provide separate input sockets. Plug the capacitor into these sockets and take the reading.

✔ **Test whether or not a diode is operational.** Digital meters with this feature have a special Diode test setting. Note that most analog meters can also test the proper operation of diodes using a low resistance scale. See the section titled "Testing diodes," later in this chapter, for details on how to do this test.

✔ **Test whether or not a transistor is operational.** Both analog and digital multimeters can perform simple testing of bipolar transistors. When using an analog meter, you can usually test the transistor in the same way that you test a diode. When using a digital meter, you test the transistor by using specially marked transistor input sockets.

✔ **Auto-zero a multimeter's reading.** For digital multimeters only, the meter automatically sets a proper zero point before taking a measurement. For analog multimeters, and some digital models, you have to first set the meter to zero. Your meter's manual outlines the precise method that you need to use.

Setting Up the Meter

Before using your meter, you must make sure that it's working properly. Any malfunction gives you incorrect testing results, and you may not even realize it.

Modern meters, especially the digital kind, require batteries. Check and replace the batteries as needed. If your meter comes with a low battery indicator or light, note when it activates and replace the meter's battery (or batteries) right away. Use only fresh alkaline batteries. Most meters aren't designed to run from rechargeable nickel-cadmium (NiCad) batteries, which deliver slightly lower voltage than their alkaline counterparts. Unless the instruction manual indicates otherwise, don't use NiCads to power your meter.

To test your multimeter, follow these steps:

1. **Turn on the meter and dial it to the Ohms (Ω) setting.**

   If the meter isn't auto-ranging, set it to low ohms.

2. **Plug both test probes into the proper connectors of the meter and then touch the ends of the two probes together, as Figure 9-8 shows you.**

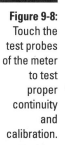

Figure 9-8:
Touch the test probes of the meter to test proper continuity and calibration.

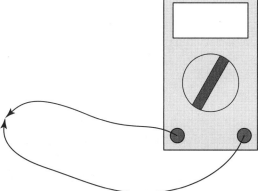

3. **The meter should read 0 (zero) ohms or very close to it.**

If your meter doesn't have an auto-zero feature, press the Adjust (or Zero Adjust) button. On analog meters, rotate the Zero Adjust knob until the needle reads 0 (zero). Keep the test probes in contact and wait a second or two for the meter to set itself to zero.

Here are some important points to keep in mind when you're testing a multimeter:

✔ Avoid touching the ends of the metal test probes with your fingers while you're performing the test. The natural resistance of your body can throw off the accuracy of the meter.

✔ Check to be sure that the test probes at the end of the test leads are clean. Dirty or corroded test probes can cause inaccurate results. Clean the probes with electronic contact cleaner, available at Radio Shack. Clean both ends of the test probes and, if necessary, the connectors on the meter.

✔ Double-check the dial setting of the meter. Make sure that you have it set to Ohms. If you don't have an auto ranging multimeter, set the range dial to the lowest Ohms setting.

You can consider the meter calibrated when it reads zero ohms with the test probes shorted together (held together so that they're touching each other). Do this test each time you use your meter, especially if you turn off the meter between tests.

Testing the resistance of good ol' water

You can use your multimeter for a simple science experiment that not only demonstrates the process of measuring resistance, but also how much crud your drinking water contains. (Yuck!) Here's how:

1. **Get two clean glasses.**

2. **Rinse out both glasses with distilled water.**

   You can get distilled water at the supermarket.

3. **Fill one glass with the distilled water and the other glass with tap water.**

4. **Set up your multimeter to measure resistance.**

   If your multimeter doesn't have auto ranging, set it to a fairly high range such as 200K ohms or higher.

5. **Strap the two test probes together at their insulating handles, using a rubber band.**

   Be sure that the metal parts of the probes don't touch.

6. **Dip the probes into the glass of distilled water. Note the reading and set the range downward, if you don't get a good reading.**

7. **Now dip the probes into the glass of tap water. Again, note the reading.**

In grade school, you may have learned that water conducts electricity. Actually, that statement isn't entirely correct. Pure water is an insulator; the minerals in the water conduct electricity. Distilled water has little mineral content, so it has a very high resistance. Depending on where you live, your tap water may contain a lot of salts and minerals, and these additions make the water more conductive. These impurities lead to water with a lower resistance that therefore better conducts electricity.

My own tests show that distilled water has a resistance of about 140KΩ; the tap water's resistance came out at about 40KΩ. Your own tests may give you higher or lower measurements for both kinds of water because of the differences in water quality and the distance between the probes.

As we talk about in the section "Okay, So What Exactly Can You Do with a Multimeter?" earlier in this chapter, many digital meters have a continuity feature that sounds a tone when the circuit you're testing reaches zero ohms. However, don't use the continuity setting for zero-adjusting the meter. The tone may sound when the meter reads a few ohms, so it doesn't give you the accuracy that you need. Recalibrate the multimeter using the Ohms setting, and not the Continuity setting, to ensure proper operation.

If you don't get any response at all from the meter when you touch the test probes together, recheck the dial setting of the meter. Nothing happens if you have the meter set to register AC or DC voltage or current. If you make sure that the meter has the right settings and it still doesn't respond, you may have faulty test leads. If necessary, repair or replace any bad test leads with a new set.

After you check the meter out, you can select the desired function (ohms, AC, DC, or current) and range and apply the probes to the circuit under test.

Five Basic Tests That You Can Make with Your Multimeter

Okay, get your meter turned on and all set up, and you're ready to make some tests. In the following sections, you learn how to conduct five of the most common tests using a multimeter.

Testing voltage

Is your circuit getting the proper voltage? You can find out with your multi-meter. You conduct voltage tests with the circuit under power. You can test the voltage at almost any point in a circuit, not just the battery connections. The procedure is simple and involves connecting the black test lead to ground, and the red test lead to a test point in the circuit that you want to check.

To perform this test

1. **Set up the meter as described in the earlier section "Setting Up the Meter."**

2. **Attach the black lead of the meter to the ground connection of the circuit.**

3. **Attach the red lead of the meter to the point in the circuit that you want to measure.**

Figure 9-9 shows an example of a multimeter testing a couple of points in a simple 555 integrated circuit (IC) timer. The top image shows the meter measuring the voltage that powers the entire circuit, and the bottom image shows the meter measuring the voltage at the output of the 555 IC. Because the output of the 555 IC is an on-or-off voltage, the reading on the multimeter alternates between zero volts and five volts.

Signals generated by circuits may change so rapidly that you can't adequately test them by using a multimeter. The multimeter can't react to the change in voltage fast enough. The proper gear for testing fast-changing signals are the logic probe and the oscilloscope. You can read about both tools in more detail in Chapter 10.

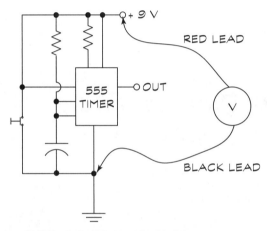

TESTING SUPPLY VOLTAGE

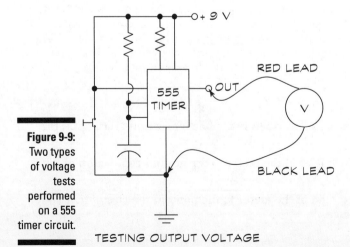

Figure 9-9:
Two types
of voltage
tests
performed
on a 555
timer circuit.

TESTING OUTPUT VOLTAGE

Testing current

As with voltage tests, you make current tests using a multimeter while you have the circuit under power. To take the basic approach, you need to connect the meter in the circuit in series with the positive supply voltage so that the meter registers the current passing through the circuit. This measurement tells you how much overall current the entire circuit draws. But remember that many digital meters are limited to testing current draw of 200 milliamps or less. *Be careful:* don't test higher current if your meter isn't equipped to do so.

You can also test current that flows through a portion of the circuit, or you can even test a single component. Figure 9-10 shows how to test the current through an LED. Make this test with the meter dialed to the Milliamperes setting.

All current measurements use this setup. You insert the meter in series with the circuit, as you can see in Figure 9-10. Connect the black lead either to ground, if testing the current draw of the entire circuit, or to the more negative side of the circuit. If you find that you get no reading at all, reverse the connections of the leads to the multimeter and try again.

After making a current test, return the meter dial to Off. This habit helps prevent damage to the meter.

Figure 9-10:
Testing
current
involves
connecting
the meter
in series
with the
circuit or
component.

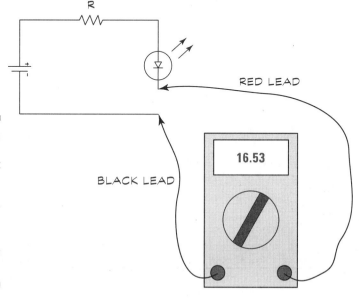

RED LEAD

BLACK LEAD

16.53

Don't blow your fuse!

Remember that hooking up a circuit or component that draws more current than your multimeter is rated to handle can lead to big problems. You run the risk of blowing the fuse in the meter, and then you have to replace the fuse before you can use the multimeter again.

Many analog and digital meters provide a separate input for testing current. If your multimeter has this input, it's usually marked as A (for amps) or mA (for milliamps). Be sure to use this input when testing current. Some multimeters have a separate input for testing higher currents, typically up to 10 amps. Typically, this input is marked as 10A.

Be sure to select the proper input before making any current measurement. Forgetting to do this step may either damage your meter (if you're unlucky) or blow a fuse (if you're lucky).

Testing wires and cables for continuity

Continuity tests whether a circuit is complete or not. We can describe continuity most clearly using a wire as the circuit:

- A *short circuit* shows that your circuit has continuity between two points of the same wire. The meter shows this state as 0 (zero) ohms.

- An *open circuit* means that your circuit doesn't have continuity. There is a break somewhere inside the wire in the circuit. The meter shows this situation as infinite ohms, which means so many ohms that the meter can't register them.

When testing a cable with many wires, you also want to determine if any of the individual wires are touching each other. When this situation happens, the wires short out. If a short happens, your circuit fails, so you want to make this test every time things go wrong.

Follow the diagram in Figure 9-11 for testing a wire by using these procedures:

- **Test for continuity in a single wire.** Connect the multimeter probes to either end of the wire. You should get a reading of 0 (zero) ohms, or very low ohms. A reading of more than just a few ohms indicates a possible open circuit.

- **Test for a short between different wires that shouldn't be electrically connected.** Connect the multimeter probes to any exposed conductor of the two wires. You should get a reading other than zero ohms. If the reading shows zero or very low ohms, it indicates a possible short circuit.

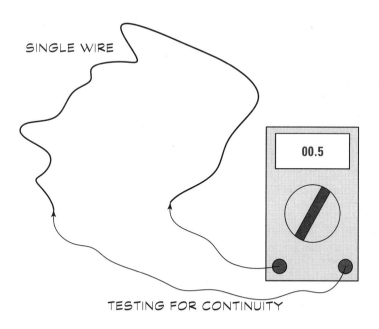

SINGLE WIRE

00.5

TESTING FOR CONTINUITY

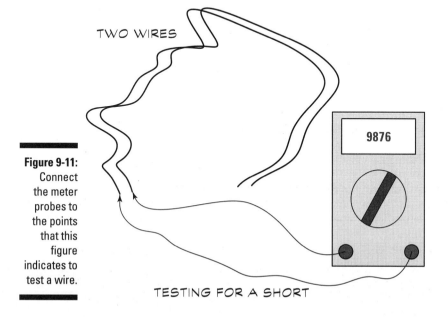

TWO WIRES

9876

Figure 9-11:
Connect
the meter
probes to
the points
that this
figure
indicates to
test a wire.

TESTING FOR A SHORT

Even wire resists the flow of electrons

Why don't you always get zero ohms when testing wire, especially a long wire? All electrical circuits have a resistance to the flow of current; the ohms measurement tests this resistance. Even short lengths of wire have a resistance, but it's usually well less than 1 ohm and so not an important test subject for continuity or shorts.

However, the longer the wire, the more the resistance, especially if the wire has a small diameter. Usually, the larger the wire, the lower its resistance per foot. Even though the meter doesn't read exactly zero ohms, you can assume proper continuity in this instance if you get a low ohms reading.

If you're testing two different wires that shouldn't be electrically connected in a circuit, you get a reading on the multimeter of infinite ohms, showing an open circuit, right? Most of the time that statement is true. But it's not always the case. Here's the reason: Even though the wires may not be directly joined, they're both connected to the circuit. This connection, whatever it is, may show a certain resistance when tested on the multimeter. So when you're looking for shorts across wires, don't be too worried if you get a reading other than infinite ohms.

Testing switches

Mechanical switches can get dirty and worn, and they can sometimes just plain break. When your switch becomes a bit worse for wear, it may no longer pass electrical current when you want it to.

Testing a wide variety of switches

Follow the diagram in Figure 9-12 to test a switch. As we discuss in Chapter 5, the most basic switch is the single-pole, single-throw, or SPST. You can readily identify such a switch by its two terminals: One acts as an inlet for the electrical current coming into the unit, and the other acts as an outlet. The switch passes or interrupts the current, depending on its position.

You call an SPST switch "single-pole" because it switches only a single part of the circuit. You use "single-throw" to describe the switch because it has only a single variation in its operating positions, on or off.

Some switches are double-pole, double-throw, or both. In a double-pole switch, the switch controls two separate circuits (say, one 12 VDC circuit and one 5 VDC circuit). In a double-throw switch, the switch may be the on-on type, or it may have a center off position, as in on-off-on. Figure 9-13 shows some variations in switch designs that you may test. Other switches may have additional poles and even three, four, or five positions, but these switches usually just come as variations on the theme (and switches with additional poles aren't common). So we don't spend extra time on them in this book.

In Chapter 5 we talk about switches and their variations in more detail.

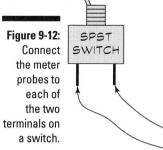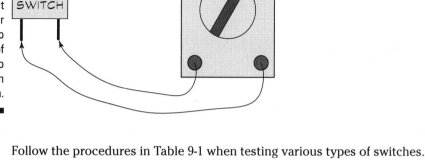

Figure 9-12: Connect the meter probes to each of the two terminals on a switch.

Follow the procedures in Table 9-1 when testing various types of switches. The physical location and function of the terminals on each switch may differ. Often, in a double-pole switch, the center terminal serves as a common; you turn the switch one way to route the current to the left terminal and turn the switch the other way to route the current to the right terminal. However, not all switches are designed this way, and only your own experimentation will help you identify the differences.

Table 9-1		Switch Types
Number of Terminals	*Type*	*Notes*
1	SPST	Metal body of the switch/second terminal. To test, connect one lead to body of switch and other lead to lone terminal.
2	SPST	To test, connect two probes of the meter to two terminals.
3	SPDT	To test, connect one lead to center terminal and other lead to one of remaining terminals. Set switch to one position and make note of the results.
4	DPST	Like three-terminal switch, but test both independent switching circuits.
6	DPDT	Like three-terminal switch, but test both positions.

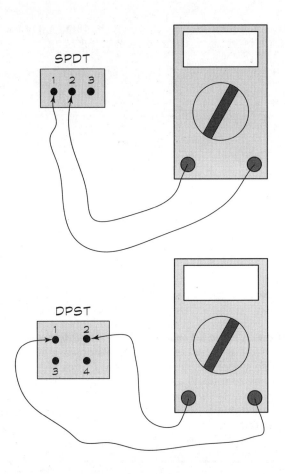

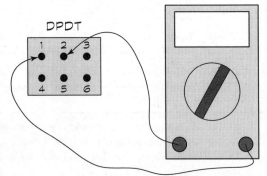

Figure 9-13: Connect the meter probes as the figure shows to test SPDT, DPST, and DPDT switches.

Some words to the wise

When testing the various types of switches, use the following tips to help you:

- ✔ With the switch in the off position, the meter should show no continuity (infinite ohms). You'll encounter this reading with both poles in a double-pole switch.

- ✔ With the switch in the on position, the meter should show continuity (zero ohms). If the meter doesn't show continuity when you place the switch in the on position, you have a fairly good indication that you're working with a bad switch.

You can most easily test switches when you work with them out of a circuit. If you have the switch wired in a circuit, the meter may not show infinite ohms when you place the switch in the off position. Instead, you may get a reading of some value other than 0 (zero) ohms. (To understand why, read the sidebar "Even wire resists the flow of electrons," earlier in this chapter.)

If you have a double-throw variety of switch, you may not have an off position. Instead, the switch has two on positions. You can test this type of switch as if it were two single-throw switches combined by making two tests rather than just one. If the switch has a center-off position, you should get a no-continuity reading for only the center position.

Testing fuses

If a circuit begins to draw too much current, it can get very hot, not only destroying itself in the process, but also possibly causing a fire. Fuses are designed to protect electronic circuitry from damage caused by excessive current flow and, more importantly, to prevent a fire if a circuit overheats. A fuse is designed to blow (or become an open circuit) when the current going through it exceeds the safe level for that fuse.

Fuses blow for reasons other than a circuit going haywire. Sometimes, they blow because of some intermittent problem, like a momentary rise (called a *spike*) in voltage from a distant (or not-so-distant!) lightning strike. When fuses blow, you need to replace them with a fuse of the same rating. You can find the fuse rating printed on the component.

To test a fuse, dial the meter to either Ohms or Continuity. Touch each end of the fuse with the meter probes, as Figure 9-14 shows. The meter should read 0 (zero) ohms. If the meter reads infinite ohms (beyond what the meter can read) it means you have a burned-out fuse and you need to replace it.

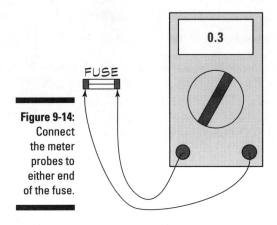

Figure 9-14:
Connect
the meter
probes to
either end
of the fuse.

FUSE

0.3

Testing Resistors, Capacitors, and Other Electronic Components

The following sections talk about the nitty-gritty reason for using a multimeter: Testing resistors, capacitors, and the other main components of a circuit.

To get more detailed information on what resistors, capacitors, and other parts that we discuss in the following sections do, see Chapter 4.

Gee, it looks all burned out!

Because the goal of testing is to determine if you have a good component to begin with, start by making first judgments based on the overall appearance of the component. In some cases, a part may be so obviously destroyed that any more testing wastes your time. You have a good sign that you're dealing with a bad electronic component when it looks burned out. If an electronic component overheats, usually as a result of soaking up too much current, it can melt or erupt. It sometimes even catches on fire! When you find a burned-out component, you need to consider why the component burned out so that you can prevent it from happening again.

Here's what you need to look for to spot damaged components:

✔ On a resistor, see if it has an obviously bulging center, with or without a distinct discoloration.

✔ On a capacitor, check for a bulge on the top or sides, with or without gooey electrolyte material seeping out. Don't worry what this gunk looks like: anything coming out of a capacitor is bad news.

✔ On a diode, transistor, or integrated circuit, look for any obvious dis-
colorations on the circuit board caused by extreme overheating of the
component.

✔ Don't overlook any component that you find in two or more pieces!
(Okay . . . duh . . .)

Avoid contact with the syrupy liquid inside an electrolytic capacitor. It's *caustic,* which means it can burn you. Wash your hands immediately with warm
water and soap if you do touch this liquid. Don't get any in your eyes! If you
do, flush your eyes out right away and seek immediate medical attention.

Of course, looks alone can be deceiving. Your component may have internal
damage, even if you don't see visual signs of burn-out. Therefore, use a visual
examination only to find obvious faults and not as a way to ultimately determine if your component has a problem. Don't assume that because everything looks okay on the outside that the component doesn't have an internal
problem.

Testing resistors

Resistors are the components that limit current through a circuit or divide
voltages in a circuit. Resistors come in a lot of values; you can find the value
marked on the body of the resistor. Sometimes, you need to verify that the
markings are accurate or that the resistor hasn't gone bad.

You can readily test resistors with a multimeter by following these steps:

1. **Set the multimeter to read ohms.**

   If you don't have an auto-ranging meter, start at a high range and work
   down.

2. **Position the test probes on either end of the resistor.**

   Be sure that your fingers don't touch the test probes or the leads of the
   resistor; if you do, you add the natural resistance of your own body into
   the reading, giving you an inaccurate result.

3. **Take the reading.**

A bad resistor can be either completely open inside, in which case you may
get a reading of infinite ohms, or it can be shorted out, in which case you get
a reading of zero ohms.

When testing a resistor, check its marked value against the reading provided
by the meter. The reading should fall within the tolerance range of the resistor. For example:

✔ If the resistor has a tolerance of 10 percent and is marked as 1K ohms, acceptable test readings fall in the range of 900 to 1,000 ohms. Tolerance is 10 percent of 1,000, or 100 ohms.

✔ If the resistor has a tolerance of one percent (you call these low-tolerance resistors *precision resistors*), acceptable test readings fall in the range of 990 to 1,010 ohms. Tolerance is one percent of 1,000, or 10 ohms.

Testing potentiometers

A *potentiometer* is a variable resistor. Like a resistor, you can test potentiometers (also called "pots") with your multimeter. As you can see in Figure 9-15, you can connect the meter to either end of the conductive material. With the multimeter applied to points 1 and 2, turning the dial shaft in one direction increases resistance. But with the meter applied to points 2 and 3, turning the dial in the other direction decreases resistance.

Figure 9-15: Connect the meter probes to the first and center, center and third, and first and third terminals of the pot.

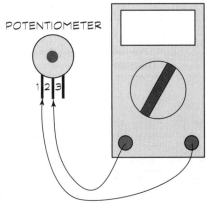

The material used for the conductive surface of the pot can take many forms, including *cermet* (a combination of ceramic, glass, and precious metals), carbon, wire, and conductive plastic. This conductive surface can break off, get dirty, or burn out. A pot sometimes goes bad because something damages the surface. As you turn the shaft of the potentiometer, use your multimeter to make note of any sudden changes in resistance, which may indicate an internal fault. If you find such a fault, you should replace the pot with a new one.

Testing diodes

A *diode* is the simplest form of semi-conductor. Diodes perform a lot of odd jobs in electronics circuits, including changing AC current to DC, blocking voltages, limiting voltage, and lighting up your life. You can test whether or not the diode functions correctly if you have a digital multimeter that has a diode-check setting.

To test a diode using a multimeter with a diode-check feature, perform these steps:

1. **Dial the meter to the diode-check setting.**

2. **Apply the test probes of the meter to the diode.**

   Observe proper polarity: Attach the red test lead to the anode (negative terminal) of the diode, and the black test lead to the cathode (the positive terminal; the cathode has a stripe so that you can identify it). Remember to avoid touching the test probes with your fingers.

3. **Observe the reading.**

4. **Reverse the probes and test again.**

Table 9-2 shows you how to interpret your test results. Although this test works for most diode types, it doesn't give a proper reading for light-emitting diodes. But you can often test light-emitting diodes visually.

Table 9-2		Display Value
1st Test	*2nd Test*	*Condition*
About 0.5*	Over range	Good
Over range	Over range	Bad — open
Zero	Zero	Bad — short

* The exact reading isn't critical, as long as it's fairly low but not zero.

Diode testing with an analog meter

If you have an analog multimeter, you can test most types of diodes by using the resistance setting and following these steps:

1. **Set the meter to a low-value resistance range.**

2. **Connect the black lead to the cathode (striped end) and the red lead to the anode.**

The multimeter should display a low resistance.

3. **Reverse the leads.**

The multimeter should display infinite resistance.

Testing capacitors

You use *capacitors* to store electrons for a short period of time. Capacitors can die an early death for a number of reasons, so use your multimeter to find out which ones you need to bury because of

- ✔ **Old age:** Certain types of capacitors, mainly those with a liquid electrolytic, can dry out over time. When they're dry, they stop working.

- ✔ **Too much voltage:** All capacitors are rated for a specific working voltage; apply voltage beyond what the capacitor is rated for and you can damage the capacitor.

- ✔ **Reversed polarity:** A polarized capacitor, which has a + or − sign marked on it, can literally blow apart if you connect it to the circuit backward.

You can check a capacitor using a multimeter that doesn't have a special capacitor-testing feature. You don't always get conclusive results, but the results you do get can help point the way to whether you should replace a component. Follow these steps to test without a capacitor-testing feature:

1. **Before testing, use an insulated bleeder jumper (see Figure 9-16) to short out the terminals of the capacitor. You can make this jumper yourself. A bleeder jumper is simply a wire with a 1 or 2 megohm resistor attached. The resistor prevents the capacitor from being shorted out, which makes it unusable.**

   This step discharges the capacitor. You need to short out the terminals because large capacitors can retain a charge for long periods of time, even after you remove power.

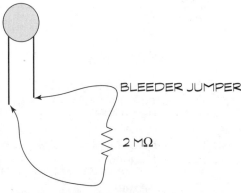

Figure 9-16: Purchase or make a bleeder jumper, useful for draining excess charge from a capacitor.

2. **Dial the meter to Ohms.**

3. **Touch the meter probes to the terminals of the capacitor. Wait a second or two and then note the reading.**

A good capacitor shows a reading of infinity when you perform this step. A reading of 0 (zero) may mean that the capacitor has shorted out. A leaky capacitor, one that is losing its ability to hold its charge, gives an ohms reading somewhere between infinity and zero.

If you are working with a polarized capacitor, connect the black lead to the – (negative) terminal of the capacitor and the red lead to the + (positive) terminal. For unpolarized capacitors, it doesn't matter how you connect the leads.

This test doesn't tell you if the capacitor is open, which can happen if the component becomes structurally damaged inside or if its dielectric (insulating material) dries out or leaks. An open capacitor will read infinite ohms. For a conclusive test, use a multimeter with a capacitor-testing function.

If your multimeter has a capacitor-testing feature, by all means use it rather than the method that we give you here. Refer to the manual that came with your meter for the exact procedure because the specifics vary from model to model. Be sure to observe proper polarity when connecting the capacitor to the test points on the meter.

You get another advantage by using a multimeter with a capacitor testing feature because the meter displays the value of the capacitor. You may find this measurement handy if you need to determine whether a capacitor falls within the tolerance range for your circuit. This feature also helps to verify the value markings on the component because not all capacitors follow the industry standard identification schemes.

Testing transistors

You can use a digital or analog multimeter to test most bipolar transistors. The test doesn't give you conclusive results, but it does provide a useful method of finding out if you have a defective transistor.

Bipolar transistors are essentially two diodes in one package, as you can see in Figure 9-17. You can therefore test the transistor by using the same methodology that we describe in the section "Testing diodes," earlier in this chapter.

Follow these steps (which assume that your multimeter has a diode-check feature) to determine if the component is good or bad:

1. **Set the meter to the diode-check setting.**

2. **Connect the red and black leads to the terminals of the transistor.**

3. **Take the reading and note the result. Refer to Table 9-3 for the results you should look for when testing a good transistor.**

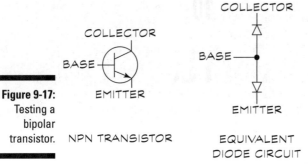

Figure 9-17:
Testing a
bipolar
transistor.

Table 9-3	Bipolar Transistor Readings
Junction Test	*Reading*
Base-emitter (BE) junction	Conduction in one direction only
Base-collector (BC) junction	Conduction in one direction only
Collector-emitter (CE) junction	No conduction in either direction

Testing with a multimeter can permanently damage some types of transistors, especially the FET (field effect transistor) type! Use this test with bipolar transistors only. Data books show these types of transistors with terminals marked as base, emitter, and collector. Schematic diagrams show the bipolar PNP and NPN resistors with either of the symbols shown here. If you're not sure whether you have a bipolar transistor, look it up in a data sheet before testing. You can find data sheets on the Internet by doing a Google or Yahoo search for the component you're interested in. Try searching by: "*2n2222 datasheet*".

If your multimeter is equipped with a transistor-checking feature, use that feature rather than the method that we give you here. Consult the manual that came with your meter for the exact procedure because it varies from one model to another.

Chapter 10

Getting Down with Logic Probes and Oscilloscopes

*I*n Chapter 9, we talk about how to use a multimeter to test for all sorts of glitches and gotchas in your electronic circuits. Your meter is the most important tool on your workbench, but don't think it's the only thing that you can use to test your electronics stuff. If you're really, really serious about electronics, you may want to get several other testing tools for your workbench.

In this chapter, we tell you about two handy test tools that you can use to make yourself a more effective electronics troubleshooter. These tools are the logic probe and the oscilloscope. Neither of the tools is a "must have," so don't rush out and buy them this afternoon. But if you start working on intermediate and advanced electronics, you may find these guys handy. Consider adding these tools to your workbench after you gain a bit of experience.

The Search for Spock: Using a Logic Probe

You use a *logic probe* (a fairly inexpensive tool), like the one in Figure 10-1, to test digital circuits. Specifically, the probe can tell you whether a signal is high or low. In digital electronics, zero volts, or very close to it, is a *low* signal. Any voltage other than zero means that you have a *high* signal. When you generate a signal that alternates between high and low very quickly you call it *pulsing*. Logic probes are great at detecting pulsing.

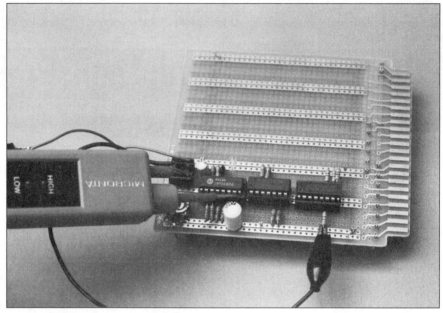

Figure 10-1:
The logic
probe is
useful when
trouble-
shooting
digital
circuits.

With few exceptions, logic circuits operate at 12 volts or less, with 5 volts as the most common. The components in the circuit define the voltage.

As you work with a logic probe keep the following "logical" tidbits in mind:

- You may see a low signal indicated by a logical 0 (zero) and a high signal indicated by a logical 1 (one). Just about every digital circuit or computer only allows the two states of 0 and 1.

- The term "logic" comes from how you combine these two states, 0 and 1, to create useful information. For example, an AND logic gate analyzes two input signals. The output of the AND gate is 1 (high) if, and only if, both inputs are 1. There are various other logic gates, including NAND, OR, NOR, and XOR. We introduce the most common of these logic gates in Chapter 1 and provide more detail in Chapter 5.

Sound, lights, action!

Although you can use a multimeter to test a digital circuit, you can use a logic probe a lot more easily. With a meter, you have to keep an eye on the readout and determine if the reading indicates low or high voltage.

With the typical logic probe, one light glows when the circuit is low and another light glows when the circuit is high (see Figure 10-2). Most logic probes also include a tone feature. The tone toggles between two states to indicate low or high. You don't have to take your eyes off the circuit; just listen to the probe as it sings to you!

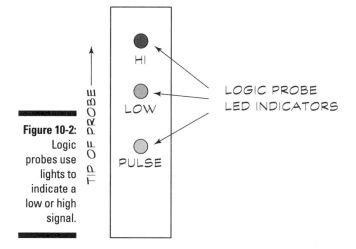

Figure 10-2: Logic probes use lights to indicate a low or high signal.

Logic probes also let you know when the circuit has no signal at all, either low or high. If the circuit has no signal, then neither probe light glows and the probe doesn't make a sound. (However, this lack of response from the probe doesn't always mean that you have a bad circuit, as we talk about in the section "What if the indicator doesn't indicate?"). With a multimeter, the lack of any signal may appear as zero volts (indicating a possible low). This difference makes a logic probe the better tool for testing digital circuits.

Logic probes also help you solve the problem of poor connections. If you have a loose wire, for example, the audible tone from the probe breaks up or crackles. You get this kind of response from the probe because it can't get a steady, reliable signal. When you hear a weak or unsteady tone, fix the connection and try it again.

Signals that are too fast (even for Superman)

Being versatile little gadgets, most logic probes can also identify a circuit where the signal is rapidly changing. This rapid signal change happens quite often in digital circuits. Figure 10-3 shows an illustration of such a changing signal, called a *square wave*. This digital signal changes, or *pulses*, between low and high. How fast it changes depends on the circuit. In some circuits, a signal changes millions of times per second.

Figure 10-3:
A typical
square
wave has its
highs and
lows.

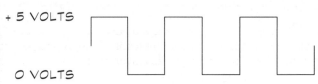

+ 5 VOLTS

0 VOLTS

Although the logic probe can't tell you how fast a signal is pulsing, for most tests you simply need to know whether the signal is pulsing at all. If you expect the signal to be pulsing, but it isn't, then you know you have a problem somewhere.

When you do need to determine the rate of pulsing, or even what the signal waveform looks like, use an oscilloscope, which you can read about in more detail in the section "Scoping Out the Oscilloscope," later in this chapter. The logic probe is a simple tool to use, and it's great for the kinds of jobs that it's designed for. But in-depth digital analysis isn't one of them.

Why don't all circuits like logic probes?

Believe it or not, some electronic circuits don't like certain pieces of test equipment. Most test gear, including the multimeter and oscilloscope, draws very little current from the circuit that you're testing. Their makers design these testing tools this way so that the tools themselves don't influence the reading. Obviously, it does no good to test a circuit if the testing tool changes the behavior of that circuit. You can't get a reliable result.

Logic probes not only draw power from the circuit, they can load down the signal line that you're testing. Some digital signals are fairly weak. The additional load of the logic probe may cause the signal to drop in voltage to a point where you can't get an accurate reading.

Although this situation doesn't come up all that often, it's a good example of why you need to be somewhat familiar with the circuit that you're testing. Just know that poking the probe into unknown territory may yield unpredictable results.

Be sure to read the manual or instruction booklet that comes with your logic probe for additional pointers, cautions, caveats, warnings, and operating tips. Though many logic probes are similar in design, slight differences can influence the types of circuits that a particular probe best works with.

Know thy circuit

To get the most use out of a logic probe, you need a wiring diagram, schematic, or service notes for the circuit that you're testing. This information helps you better determine the source when you discover problems.

You also need to have some documentation for the circuit handy because you have to be careful about where you put the logic probe. A logic probe receives its power from the circuit that you're testing. To use the probe, you must first connect its power leads to the positive and ground connections of the circuit that you're troubleshooting. You're not supposed to operate most logic probes at more than 15 volts, so you have to know where to tap into the circuit for the power. If you connect the probe to a spot with a high voltage, you run the risk of permanently damaging the probe, the circuit you're testing, or both. *If you don't know the voltage level of a particular circuit, first test it with a multimeter.*

Putting the Logic Probe to Work

No doubt, you're dying to see the logic probe in action. In the following sections, we run down some safety issues that you need to be aware of before you start, take you through the steps of using a logic probe to test a circuit, and tell you just what the readings you get from the probe may mean.

Observe the usual safety precautions, please

The same safety precautions that you use with a multimeter apply when you use a logic probe, only more so. We won't repeat those precautions here, but you should take a quick look at Chapter 9 before you actually start using a logic probe.

Safety is even more important with a logic probe than with a multimeter because the logic probe is an active-circuit tester. You have to turn the circuit on in order to test it. This requirement is not always true of the multimeter, with which you can conduct certain tests, such as continuity (testing whether a circuit is complete), without applying any power to the circuit.

Take special care if the circuit that you're testing runs off AC power and you need to expose the power supply components to perform the test. You may find yourself in this situation, for instance, if you're trying to figure out why your VCR is on the fritz. Always consider that you may expose dangerously high voltages when you remove the cover from any AC-operated equipment. If you're working close to equipment that conducts these voltages, cover the equipment with insulating plastic to prevent accidental shock.

Connecting the probe to the circuit

The logic probe has four connections, as you can see in Figure 10-4. The red and black leads use alligator clips so that you can securely attach them to ground and the power supply of the circuit that you're testing.

Be sure to first determine if the supply voltage of the circuit falls within the acceptable range for the logic probe. Most probes work with a minimum supply voltage of about 3 volts and a maximum of no more than 15 volts (sometimes more, sometimes less). For the exact voltage range of your logic probe, check the manufacturer's instruction booklet.

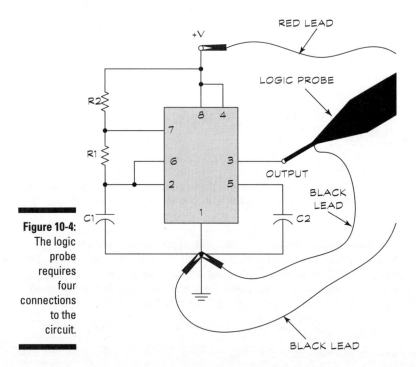

Figure 10-4:
The logic probe requires four connections to the circuit.

You need to make these four connections:

- ✔ You clip the black power lead to circuit ground.

- ✔ Clip the red power lead to the circuit voltage supply. Be sure that this supply doesn't exceed about 15 volts, or you can damage the logic probe.

- ✔ You connect a second black ground lead to circuit ground. This separate ground is important; if you fail to solidly connect the probe to circuit ground, the probe may fail to work, or it may yield erratic results.

- ✔ Place the tip of the probe against the part of the circuit that you're testing.

When you've made these connections, observe the probe's reaction. Indicator lights and audible tones (on most logic probes) help you determine the logic level at the test point:

- ✔ **Low indicator (accompanied by the low buzz tone):** This reaction tells you that the test point has a logic low (at or about 0 volts).

- ✔ **High indicator (accompanied by the high buzz tone):** This result indicates that the test point has logic high (usually at or about 5 volts).

- ✔ **Quickly toggling Low and High indicator:** This reaction means that the logic signal is *pulsing* (changing between low and high at a fast pace). *Note:* Most logic probes have a separate indicator that tells you when a circuit is pulsing.

- ✔ **No indicator:** If you get nothing from your probe, the test point has no discernable high, low, or pulsing signal.

What if the indicator doesn't indicate?

You may find logic circuits bewildering beasts, especially if you're new to working with them. In some instances, the output of a logic circuit may yield no indication of a signal. When you get this indication from your probe it doesn't necessarily mean that you have a faulty circuit. (But remember that, in many cases, no signal means that your circuit does have a problem.) When the logic probe gives no indication of a signal, the lack of signal can mean that you have either a bad circuit or an incorrectly connected logic probe, or both.

When you're trying to figure out why the probe doesn't react to your circuit, having a wiring diagram or some type of schematic for that circuit comes in handy.

You can help rule out the chance that you've incorrectly connected the probe by conducting this quick test:

1. **Touch the test lead of the probe to the power supply of the circuit. The probe should indicate a high value.**

2. **Next, touch the test lead of the probe to the ground of the circuit. Now, the probe should indicate a low value.**

3. **If either or both tests fail, examine the connection of the probe to the circuit and make corrections as necessary.**

Assuming that you have the probe connected properly, you can now use the logic probe at additional test points in the circuit.

In the section "The Search for Spock: Using a Logic Probe," earlier in this chapter, we tell you that a logic circuit has only two possible outputs: low or high. Although that's technically true, some kinds of integrated logic circuits have a third state, called Hi-Z or high-impedance. The reasons why this third state exists go a little beyond the scope of this book, but in general, *Hi-Z* lets you connect a lot of outputs directly together, with only one being active (or enabled) at a time. The remaining outputs are set to their Hi-Z state, which makes them essentially invisible to the enabled output. The circuit only engages one output, either low or high, at any one time. The other outputs are put to sleep in the Hi-Z state and get activated in their own due time.

Scoping Out the Oscilloscope

A true electronics gearhead poses for the high school alumni newsletter picture standing next to an oscilloscope. The scope is something of a badge of honor. If you have one, let alone know how to use it, then everyone assumes that you're an electronics guru. Some things just look cool.

Though a little on the expensive side, the oscilloscope is the tool that any die-hard electronics tech needs. For the average amateur electronics hobbyist working at home or in school, the oscilloscope is a nice tool to have around, but if you're less than obsessed, you don't absolutely need it. So unless you just *have* to look like a gearhead, you can get by without one . . . for a little while, at least.

Although not everyone owns an oscilloscope, and not all electronics projects require one, it still makes sense to introduce them to you and provide the basics about how they work to complete your basic electronics education.

So, exactly what does it do?

The job of the oscilloscope is to visually represent an electrical signal, either alternating current (AC) or direct current (DC). An oscilloscope shows variations in voltage as a bright line drawn on the display, as you can see in Figure 10-5.

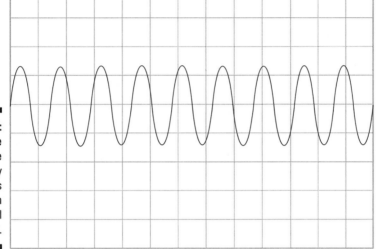

Figure 10-5:
The oscilloscope graphically displays changes in an electrical signal.

DC voltages appear as straight lines; their position vertically indicates the voltage value. AC voltages appear as undulating lines, also called *waveforms.* An oscilloscope shows both the AC signal voltage and its frequency. Pretty cool stuff.

Figure 10-6 shows a fairly typical bench top oscilloscope identifying common dials, jobs, and other controls. We get to what these features all mean in the section "The Ins and Outs of Using an Oscilloscope," later in this chapter.

The oscilloscope screen displays a grid; the X (horizontal) axis represents time, and the Y (vertical) axis represents volts. Count the number of divisions on the screen's grid to determine the voltage and (if you're using an AC or digital signal) the variation of the signal over time.

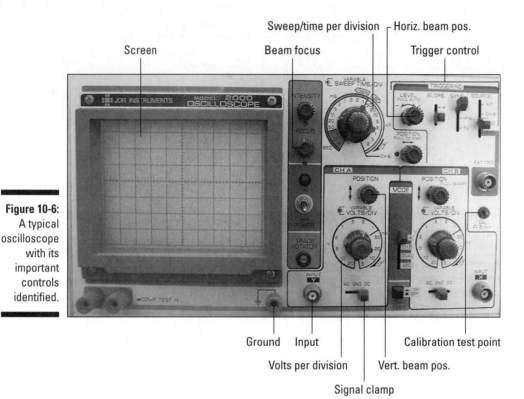

Screen   Beam focus   Sweep/time per division   Horiz. beam pos.   Trigger control

Figure 10-6:
A typical
oscilloscope
with its
important
controls
identified.

Ground   Input   Calibration test point

Volts per division   Vert. beam pos.

Signal clamp

Sticking to common oscilloscope features

For the purposes of this chapter, we assume that you don't have wads of
money in your pocket to buy the latest whiz-bang scope. Here's what we
assume you probably have:

- ✔ You may already have an oscilloscope, maybe even an old relic, but
don't yet know how to use it. Time to dust it off, plug it in, and fire it up!

- ✔ You have access to a scope through school or work. Maybe you can
arrange to borrow it in the school/work lab or even to take it home to
your own bench for those times when an oscilloscope is the tool that
you absolutely have to have to solve a problem.

- ✔ You find a great bargain on a used oscilloscope on eBay, and you're willing to take a chance. (You can often find a fairly nice used model for
under $100.)

With these assumptions in mind, this chapter limits itself to the most
common features found on oscilloscopes. We skip the very high-end stuff and
cover the features common to just about every scope made since 1970.

Oscilloscopes are fairly complicated pieces of equipment, and to thoroughly understand their proper use, read the instruction manual that comes with the scope or a book dedicated to the subject. This chapter gives you just a quick overview to get you started. Visit one of the electronics sites mentioned in the Appendix; many provide tutorials on using a scope.

Bench, handheld, or PC-based?

You can still find the old model of oscilloscope, with its dials, knobs, switches, and a cathode ray tube (CRT). In fact, professional electronics field technicians still prefer bench CRT scopes. But these days, you can choose from several types, and each type has its advantages and disadvantages. We take a moment to review them in the following sections.

Bench scope

The bench oscilloscope, like the one in Figure 10-7, provides a bright and clear image on its glowing green screen. Even on the latest models, you can get your hands on all the basic functions and controls with front-panel switches and dials. This setup allows you to quickly select the operating modes of the scope without meandering through a series of on-screen programming menus.

If you're looking for a used oscilloscope, you're most likely to find the bench variety. They've been making them for decades, but the newer models include computer interfaces and other advanced features.

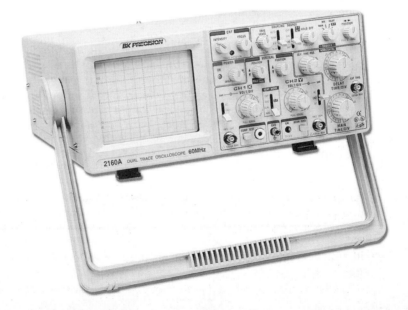

Figure 10-7: A bench scope sits on your worktable and is completely self-contained.

Some enhanced features you should know about

Oscilloscopes have improved greatly over the years, with many added features and capabilities. Although you don't absolutely need any of the following features for routine testing, you may find them handy as you gain experience. Among the most useful features are

✔ **Delayed sweep:** When analyzing a small portion of a long, complex signal, this feature helps because you can zoom into just a portion of the signal and examine it. This is ideal when you work with television signals.

✔ **Digital storage:** This feature records signals in computerized memory for later recall. After you have it in the memory, you can expand the signal and analyze specific portions; again, helpful in television work. Digital storage also lets you compare signals, even if you take the measurements at different times.

As you may expect, these features can add to the cost of the scope. Balance the extra cost against the usefulness of the features.

Handheld scope

For on-the-go work, nothing beats a portable handheld oscilloscope. Looking a bit like a Star Trek tri-corder, the handheld scope provides all the basic functions of a scope, but in a battery-operated, palm-sized tool. The screen is a liquid crystal display, and although smaller than the screen on the average bench scope, it's still functional and readable.

You may find handheld oscilloscopes very handy (no pun intended!) if you're on the go, but they don't usually offer all the advanced features of the better bench oscilloscopes. If you need the latest and greatest, rely on a handheld model only when portability is a must.

PC-based scope

The PC-based oscilloscope doesn't have a screen of its own. Instead, it uses your personal computer (either desktop or laptop) to store and display the electrical signals that you measure. Most PC-based scopes are self-contained in a small, external module. The scope connects to the desktop PC or laptop through a parallel, serial, or USB port. A few manufacturers have designed PC-based oscilloscopes that you can install inside your computer. These plug into one of the available expansion slots inside the chassis of your desktop PC.

Many PC-based oscilloscopes cost less than a similarly equipped bench scope. They also take advantage of the features built into your PC, such as disk storage and printing. Obviously, the PC-based scope has one big disadvantage: You have to have access to a PC, either permanently plopped on your electronics bench or temporarily brought in (as with a laptop) when you need it.

Understanding oscilloscope bandwidth and resolution

You should know about a couple of notable specifications for oscilloscopes. One of the most important specs is bandwidth. *Bandwidth* is the highest frequency signal that you can reliably test with your oscilloscope, measured in megahertz (MHz). PC-based scope probes tend to have the lowest bandwidth, usually about 5-10 MHz. This bandwidth works just fine for many tasks, including working with hobby circuits and even servicing VCRs and audio equipment.

The average bandwidth of a low-cost bench scope falls in the 20-35 MHz range. This range does the job for all but the most demanding applications. Specialized troubleshooting and repair, such as work on computers and ultra-high-frequency radio gear, may require bandwidths exceeding 100 MHz. But remember that the price of an oscilloscope goes up considerably as the bandwidth gets higher.

Another important specification is resolution. The *resolution* of the scope has to do with its accuracy. The X (horizontal) axis on an oscilloscope displays time, and the Y (vertical) axis displays voltage. The horizontal amplifier indicates the X-axis resolution. Most scopes generally have a resolution of 0.5 microseconds (millionths of a second) or faster. You can adjust the sweep time so that you can test signal events that occur over a longer time period, usually as long as a half a second to a second. Note that the screen can display signal events faster than 0.5 microseconds, but such a small signal may appear as a fleeting glitch or voltage spike.

The sensitivity of an oscilloscope indicates the Y-axis voltage per division. The low-voltage sensitivity of most average-priced scopes is about 5 mV (millivolts, or thousandths of a volt) to 5 volts. You turn a dial to set the sensitivity that you want. When you set the dial to 5 mV, each mark on the face of the scope tube represents a difference of 5 mV. Voltage levels lower than 5 mV may appear, but you can't accurately measure them. Most scopes show very low voltage levels (microvolt range) as a slight ripple.

The ins and outs of using an oscilloscope

Although an oscilloscope can do some pretty cool things, you only have to perform a couple of steps to actually use one.

Here's a quick rundown of the steps that you perform to measure the voltage of a DC signal with an oscilloscope:

1. **Attach a test probe to the scope input.**

   Note: Some scopes have several inputs, called channels; we assume you're dealing with just one input for now.

2. **Adjust the Volts Per Division control to set the amplitude or voltage range.**

   For example, if the voltage you're testing is 0-5 volts, use the 1 volt per division range. With that setting, each volt corresponds to one tick mark on the screen of the scope.

3. **Adjust the Sweep/Time Per Division control to set the time slice of the signal.**

   The *time slice* is the duration of the part of the signal that's shown on the scope. A shorter time slice shows only a brief portion of the signal, whereas a longer time slice shows you more of it.

   If you're testing a DC signal, you don't need this control because the signal doesn't change (much) over time. You can choose a medium-range setting to ensure consistent readings, such as 1 millisecond per division (a millisecond is one one-thousandth of a second).

4. **Select the signal type, either AC or DC, and the input channel.**

   Note that you don't get an input channel selector if you buy a single-channel oscilloscope.

5. **Most scopes have a trigger switch. If yours does, set it to Auto.**

6. **When you've set up the oscilloscope properly, connect the test probe to the signal that you want to test.**

7. **Connect the ground of the probe to the ground of the circuit.**

8. **Connect the probe itself to the circuit point that you want to test (you can see this setup in Figure 10-8).**

9. **Read the waveform displayed on the screen.**

   Unless your scope has a direct read-out function that displays voltages on the screen, you need to correlate what you're seeing with the settings of the s.

If you're testing a low-voltage AC or pulsing digital signal, set the Sweep/Time Per Division control so that you can adequately see each cycle of the signal. Don't worry . . . you can experiment with the Sweep/Time Per Division control until the signal looks the way that you want it to.

Do not test AC voltage coming from a wall outlet using an oscilloscope, unless you first take special precautions. The manual that came with your scope should outline those precautions. We assume you are using your oscilloscope only to test low-voltage DC circuits, and low-voltage AC signals, such as those from a microphone. If you connect your scope directly to 117 VAC from a wall outlet you can injure both you and your scope!

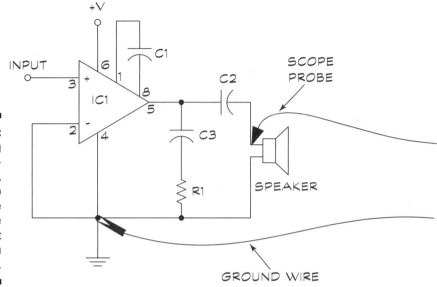

Figure 10-8:
When using
an oscillo-
scope,
touch the tip
of the scope
probe to the
circuit point
that you
want to test.

What all the wiggly lines mean

Oscilloscopes give you a visual representation of an electrical signal. The vertical axis indicates the amount of voltage (also called *amplitude*), and the horizontal axis represents time. Oscilloscopes always sweep left to right, so you
read the timeline of the signal from left to right, just as you read a line in a book.

The signal that you observe on the oscilloscope is a *waveform.* Some waveforms are simple, some are complex. (We introduce the concept of waveforms
and different signal types in Chapter 1.) Figure 10-9 shows the four most
common waveforms that you encounter in electronics and what these waveforms look like on an oscilloscope screen:

✔ **DC (direct current) waveform:** A flat, straight line, like the one that you
   see here. A DC waveform's amplitude, which is the voltage reading, is
   critical.

✔ **AC (alternating current) waveform:** This waveform undulates over
   time. The most common waveform is a sine wave (see Chapter 1 for
   more about sine waves). AC waveforms vary in frequency. Some move
   quite slowly, such as 60 Hz (60 cycles per second), the frequency of
   house current in North America. Or AC waveforms can move very
   quickly, on the order of several million or billion Hertz.

✔ **Digital waveform:** A DC signal that varies between no volts (low) to some pre-determined voltage (high). The digital circuitry interprets the timing and spacing of the low and high marks. When you plug in a digital camera to your computer, the computer copies the pictures stored in the camera to its hard drive by using such a waveform. The waveform changes very quickly so that you can transmit the data in a short period of time.

✔ **Pulse waveform:** This waveform shows a sudden change between a signal's low and high states. Most pulse waveforms are digital and usually serve as a timing mark, like the starter's gun in a 440 yard dash. When the gun goes off (the pulse) other parts of the circuit react and generate even more signals.

DC WAVEFORM

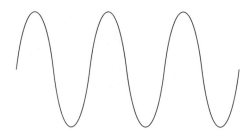

AC WAVEFORM

DIGITAL WAVEFORM

Figure 10-9:
Four
common
waveform
forms

PULSE WAVEFORM

So, When Do I Use an Oscilloscope?

When you're testing voltage levels, you can often use multimeters and oscilloscopes interchangeably. The choice of which tool you use is yours, though for routine testing procedures, you may find the multimeter a little easier. In general, you may opt to use an oscilloscope for

- ✔ **Visually determining if an AC or digital signal has the proper timing.** For example, you often need this test when you troubleshoot radio and television equipment. The service manuals and schematics for these devices often show the expected oscilloscope waveform at various points in the circuit so that you can compare. Very handy!

- ✔ **Testing pulsating signals that change too rapidly for a logic probe to detect.** Generally these are signals that change faster than about five million times a second (5 MHz).

- ✔ **Visually testing the relationship between two input signals,** when using an oscilloscope with two input channels. You may need to do this test when you work with some digital circuits, for example. One signal may trigger the circuit to generate yet another signal. In fact, this is quite common. Being able to see both signals together helps you determine whether the circuit is working as it should.

- ✔ **Testing voltages,** if the scope is handy; but you can use your multimeter for testing voltages, too.

Rather than whipping out your oscilloscope for every test, you're better off using a multimeter for the following:

- ✔ Testing the resistance of a circuit
- ✔ Determining if a wire or other circuit is shorted (0 ohms resistance) or open (infinite ohms resistance)
- ✔ Measuring current
- ✔ Testing voltages and various components, such as capacitors and transistors

Putting the Oscilloscope to Work: Testing, 1-2-3!

So if you've been reading along in this chapter, you now know a little bit about what an oscilloscope is for and what it does. In the following sections, we show you how to do a couple of quick tests. These tests demonstrate how you use a scope for a variety of simple chores. After you work through these tests, you're well on your way to becoming a master scope user.

Basic setup and initial testing

Before you use your oscilloscope for any actual testing, set its controls to a normal or neutral setting. You then calibrate the scope, using its built-in test point, so that you're sure it's working correctly.

Here are the steps for setting up your scope. Refer to Figure 10-6, earlier in this chapter, to reference the various knobs and buttons on your scope as you go through these steps. Remember that your oscilloscope may look a bit different, and its knobs and controls may have slightly different names.

1. **Turn the scope on.**

   If it's the CRT bench top variety, allow time for the tube to warm up. You may or may not see a dot or line on the screen.

2. **Set the Sweep/Time Per Division knob to 1 millisecond.**

   This setting is a good middle value for initial calibration.

3. **Set the Volts Per Division knob to 0.5 volts.**

   This setting is also a good middle value to use for initial calibration when testing low-voltage DC circuits.

4. **Set the Trigger Level control to Automatic (or midway, if it doesn't have an Automatic setting). Select AC Sync and Internal Sweep.**

5. **Select the Auto setting for both Horizontal Position and Vertical Position; or you can crank the knobs to their midpoint if your scope doesn't have an Auto setting.**

6. **Connect a test probe to the input.**

   If your scope has multiple channels (sometimes called inputs), use Channel A.

7. **Select Gnd (Ground) for the Signal Clamp, if your scope has this control.**

   On some scopes, this control may be called Signal Coupling.

8. **Connect the ground clip of the test probe to the designated ground connection on the scope (see Figure 10-10).**

   If your oscilloscope doesn't have a designated ground connection, clip the lead to any exposed metal point, such as the head of a screw.

9. **If your scope has a Signal Clamp switch, attach the center of the test probe to the calibration test point. If your scope lacks a Signal Clamp switch, attach the center of the test probe to the ground point.**

Figure 10-10:
Connect the ground of the probe to the ground connection on the scope.

10. **Adjust the Vertical Position knob until the beam sits on the first division on the screen (Figure 10-11).**

11. **Adjust the Horizontal Position knob until the beam is more or less centered on the screen.**

    You don't need to worry about making this setting exact.

12. **If your scope has a Signal Clamp switch, set it to DC. If you don't have a Signal Clamp switch, move the test probe from its ground connection to the calibration test point.**

Many oscilloscopes use a test signal that appears as a relatively low-frequency square wave. (Forgot what a square wave looks like? Review Chapter 1 for details.) Consult the manual that came with your scope to see what voltage and frequency your scope produces with its built-in test calibration circuit.

For example, say that the signal should be 0.5 volts peak-to-peak (indicated as 0.5v p-p) at 1000 Hz. Because you set the Volts per Division knob to 0.5 volts and the test signal has amplitude of 0.5 volts, the waveform spans one division on the screen.

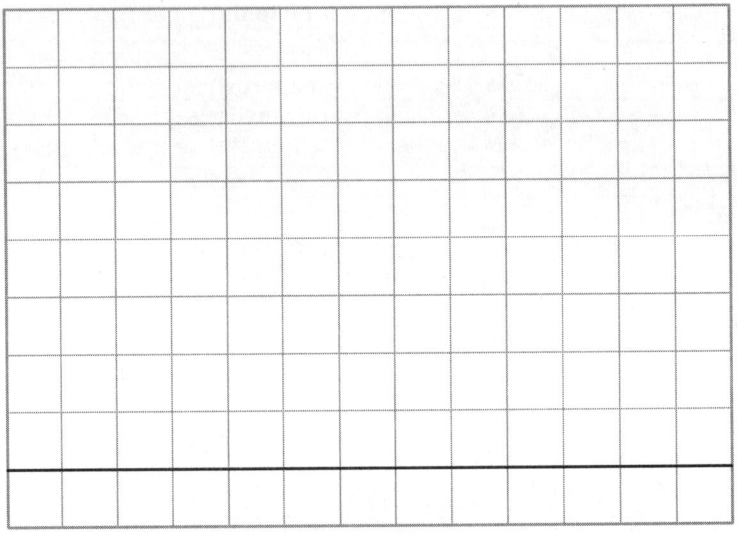

Figure 10-11:
Adjust the
Vertical
Position
knob so that
the beam
sits on the
bottom of
the grid.

By decreasing the Volts Per Division setting, you can make the waveform larger. Do this adjustment when you need more accuracy. For example, if you set the Volts Per Division knob to 0.1 volts, a 0.5 volt test signal spans five divisions.

Does your battery have any juice?

Admit it. You have a drawer full of spare batteries, and you sure would like to know just how much voltage the darn things have to give, right? A rudimentary test that you can perform using the oscilloscope is measuring voltage. A battery produces a DC voltage, so the sweep setting on the scope is irrelevant in this test. You just want to know what voltage the scope displays on the screen.

For this demonstration, you test the voltage of a 9-volt battery: So, first, dig around in your drawer and pull out a 9-volt battery; then, follow these steps:

1. **Follow the basic setup and initial testing procedure outlined in the previous section.**

2. **Set the Volts Per Division knob to 2 volts.**

3. **Attach the ground clip of the test probe to the – (negative) terminal of the battery.**

4. **Attach the center of the test probe to the + (positive) terminal of the battery.**

   The line on the screen should fall approximately mid-way between the fourth and fifth divisions. Because you have the Volts per Division knob set at 2 volts, this line placement indicates that your battery has 9 volts, or 4.5 divisions times 2.

Dissecting your radio to display an audio waveform

You can have a lot of fun with this test because you get to take apart a gadget to run the test. Oscilloscopes can visually represent the AC waveform, which is the electrical signal that drives a speaker. The AC waveform is complex because it's made up of constantly changing frequencies. These changing frequencies are what you hear as singing, talking, or the sound of musical instruments.

For this test, pry off the back of an ordinary battery-powered radio so that you can reach the two terminals on the speaker. Then, follow these steps:

1. **Follow the setup-and-testing procedure outlined in the section "Basic setup and initial testing," earlier in this chapter.**
2. **Set the Volts Per Division knob to 1 volt.**
3. **Set the Sweep/Time Per Division knob to 100 microseconds.**
4. **Attach the ground clip of the test probe to one of the speaker terminals.**
5. **Attach the center of the test probe to the other speaker terminal.**
6. **Turn on the radio and watch the display.**
7. **If at first you don't get much of a reading, try, try again by decreasing the Volts Per Division setting.**

Here are some things to watch for when you perform this test:

- ✔ **The amplitude of the waveform increases and decreases as you change the volume on the radio.** This change happens because the volume control alters the signal voltage that you apply to the speaker.
- ✔ **By turning the Sweep/Time -Per Division knob, you can see finer detail's of the signal.** A slower sweep of, say, 0.1 milliseconds displays frequencies of up to 1000 Hz per division. A faster sweep of, say, 100 microseconds displays frequencies of up to 10,000 Hz per division.

If you have access to a signal test generator that can produce a single tone, you can use this same technique to take a look at its waveform. Rather than a mish-mash of squiggly lines, you see a distinct sine wave.

Testing the frequency of an AC circuit

You can determine the frequency of an AC signal by using an oscilloscope. Now, although you can plug the test probe of the scope into a wall socket to display the 60 Hz (50 Hz in some parts of the world) alternating current coming out of it, DON'T! This poses a significant safety hazard so don't even think about it. Instead, test the frequency of your household current indirectly (and more safely), using a phototransistor.

For this test, you need a phototransistor (not a photodiode or photoresistor, but a phototransistor, which is a light-dependent transistor), and a 10K resistor (head over to Chapters 4 and 5 if you need a refresher on these parts). Connect the phototransistor and resistor to a 9-volt battery, like the setup in Figure 10-12. Grab a lamp outfitted with an incandescent bulb, and you're ready to go!

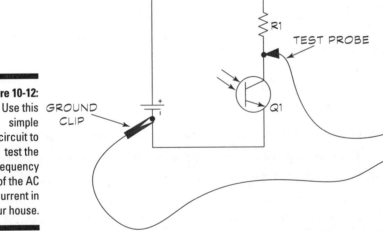

Figure 10-12:
Use this simple circuit to test the frequency of the AC current in your house.

1. **Follow the setup-and-testing procedure outlined in the section "Basic setup and initial testing," earlier in this chapter.**

2. **Set the Volts Per Division knob to 1 volt.**

3. **Set the Sweep/Time Per Division knob to 10 milliseconds.**

4. **Attach the ground clip of the test probe to the – (negative) terminal of the battery.**

5. **Attach the center of the test probe to the point where you have connected the phototransistor and resistor.**

6. **Turn on the light and note the ripple in the waveform.**

This ripple is the AC current pulsing through the incandescent lamp. For best results, don't shine the lamp directly into the phototransistor. This direct exposure may swamp the transistor, and you can't see any signal change. Direct the light away until you can see a sine wave. Adjust the Volts Per Division knob until you get a decent reading.

Have you been staring at the light, trying to see the changes in brightness as the AC current pulses? You can't see the lamp getting alternately brighter or dimmer with the naked eye because of a phenomenon called persistence of vision. But the phototransistor acts much faster than your eyes: It can detect very quick changes in light.

An oscilloscope shows you the period of an AC signal, not its frequency. You have to do some math to convert one into the other. A 60 Hz signal has a period of 0.0166 seconds, which you can determine with a bit of simple math:

$$\frac{1}{\text{Frequency}} = \text{Period (in seconds)}$$

You convert time period to frequency by changing the math around a little:

$$\frac{1}{\text{Period}} = \text{Frequency (in Hertz)}$$

Forget for the moment that you already know that the frequency of the AC current that the scope senses through the phototransistor is 60 Hz. To determine the frequency just by reading the scope, you first measure the distance from crest-to-crest of the waveform. You then do the math that we just walked you through. Assuming that you set the Sweep/Time Per Division knob to 10 milliseconds, each crest-to-trough transition of the AC signal spans about 1.6 divisions.

$$\frac{1}{1.6} = 0.625$$

Because you set the Sweep/Time Per Division knob to 10 milliseconds, you have to divide the result by 0.01 (10 milliseconds). You end up with 62.5, close enough to 60 Hz for our purposes.

But wait! When you test the output of the phototransistor, the waveform spans only about 0.8 of a division. What gives? The phototransistor actually registers 120 flashes of light per second because the light pulses each time that the AC current goes positive or negative. In this case, because we're testing the AC waveform indirectly, you need to cut the result in half.

$$\frac{1}{0.8 \ (\text{the measured period})} = 125$$

Divide 125 by 2, and you get (approximately) 60 Hz.

You may have noticed that an oscilloscope provides only an approximation of the frequency of a signal. If you need something more accurate, you need a frequency counter. These babies use a digital readout to display signal frequency, and they're accurate to within one in several hundred thousand Hertz. Check out Chapter 16 for more about frequency counters.

Part V

A Plethora of Projects

In this part . . .

In the three chapters that follow you discover how to build your own circuits using a nifty little device called a solderless breadboard. You just plug in parts and you've got a circuit! Changes are easy to make, and when you're done experimenting, you can produce a permanent soldered circuit board. The chapters in this part tell you how to do that, too.

And it gets even better. In the pages that follow you also uncover the magic and power of something called a microcontroller that allows you to program an electronic gadget to do your bidding. Finally, you can move on to Chapter 14 and follow along with a dozen fun projects you can build in 30 minutes or less. There are even a couple of robot projects in Chapter 15 for making your own personal robot pal.

Chapter 11

Creating Your Own Breadboard Circuit

*Y*ou may think that you'd get funny looks if you ask for a "breadboard" at your local super-duper electronics-parts mart. After all, what does bread have to do with electronic gizmos? But no, you can expect to get a smile of appreciation for your do-it-yourself spirit, and the friendly sales staff will point you to the make-your-own-circuits aisle. In that aisle, you find a bunch of funny-looking white square or rectangular plastic thingies with more holes than Swiss cheese.

These little critters are circuit breadboards, and you use them to experiment with all kinds of electronic ideas without having to warm up your soldering iron. You use them to create a sort of rough draft of a circuit board that you can play with before you go to the trouble of creating a finished, printed circuit board. It's easy to fix mistakes at the breadboarding stage.

In this chapter, we tell you all about circuit breadboards and how to use them. They're pretty simple, actually. You don't need a degree in engineering to use them, but we do have some tips and suggestions that you don't want to miss.

We also include a couple of circuit board construction techniques in this chapter, including point-to-point wiring and something called wire wrapping. You can use these techniques when you're ready to make a permanent circuit.

For now, though, you just play!

Taking a Look at Solderless Breadboards

Circuit breadboards, also called prototyping boards or solderless breadboards come in all shapes, styles, and sizes, but they all serve the same function: They have columns of holes that little slivers of metal connect electrically. You plug in components — resistors, capacitors, diodes, transistors, integrated circuits . . . you name it — and then string the wires to build your circuit. When you're confident that the circuit works, you can use one of the many construction techniques available to you. (We talk about some of your construction options in Chapter 12.)

Don't skip the step of first testing all the circuits that you plan to build on a solderless breadboard. Often, you can improve on the performance of the circuit just by tweaking a few component values. You can easily do these changes by simply removing one component on the breadboard — without having to unsolder and resolder — and exchanging it for another.

The following sections tell you everything you ever wanted to know about solderless breadboards.

Solderless breadboards, inside and out

A solderless breadboard, like the basic model in Figure 11-1, consists of a series of square holes, and inside the holes are rows and rows of metal strips. The metal strips, which are made of a flexible material, have been bent to make a channel. You slide wire into the holes and securely connect it inside the metal channel.

You call the metal channels inside the breadboard *contacts*. Each column in each row connects electrically. Each column connects together five holes, all also electrically joined. With this setup, you can connect components and wires just by plugging them into the right holes on the board. See Figure 11-2 for a visual representation of how the columns in a breadboard connect electrically.

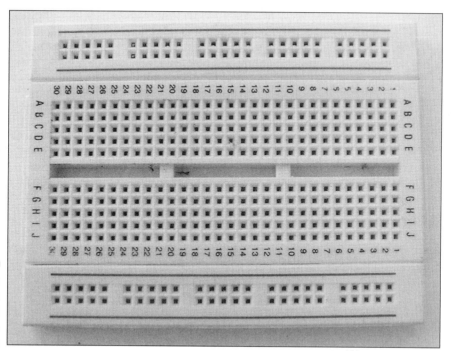

Figure 11-1:
A basic
solderless
breadboard.

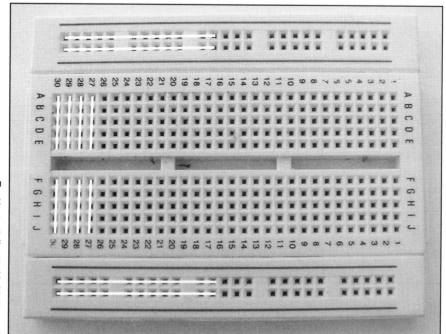

Figure 11-2:
Bread-
boards
consist of
columns
that
connect
electrically
inside.

Note also the long rows of contacts at the very top and bottom of the solderless breadboard. These rows help you make convenient connections to power and ground. Most boards have two rows at the top and bottom. In some boards, the two rows connect electrically, but on others, each row is electrically isolated from the other. Don't assume! Use your multimeter to check by sticking a jumper wire in each hole and then touching one probe to one wire and another probe to the other wire. If you get a low ohm reading, the two rows are connected together. If you get an infinite ohms reading, they're not connected. See Chaper 9 for more about testing things with your multimeter.

The holes are spaced $\frac{1}{10}$ of an inch apart (0.100 inch), a size just right for integrated circuits, most transistors, and discrete components, such as capacitors and resistors. You just plug in ICs, resistors, capacitors, transistors, and 20- or 22-gauge wire in the proper contact holes to create your circuit.

Solderless breadboards are cheap . . . er, inexpensive

Author Gordon McComb here to relate a little personal story: For years, and I mean *years,* I got by with a single solderless breadboard. Every time I wanted to try a new circuit, I had to disassemble the old one to make room for my latest and greatest idea. Sometimes I wasn't completely finished with the old circuit, so I had to rebuild it from scratch. That meant taking apart my new circuit and reconstructing the old one. Needless to say, this process was very time consuming (and more than a little annoying).

In hindsight, I can see how dumb my single-breadboard lifestyle was. Basic solderless breadboards aren't all that expensive. Most cost under $10. You make things a lot easier on yourself by simply having several solderless breadboards in your toolkit and building different circuits on them. You can leave a circuit on the board until you're done with it or until you run out of boards for your new projects.

You can also use multiple breadboards to build circuits bit by bit. You essentially build each part of the circuit as a module. This approach lets you experiment with different sections of a more complicated circuit, perfecting each section before going on to the next. You can then use a couple of wires to easily connect the modular breadboards together.

To make a really sophisticated experimenter's station, buy some wide Velcro® strips and attach the strips onto a piece of 12 x 12-inch wood. You can get wood already cut to this size at most hobby or craft stores. Then, stick a strip of the mating piece of the Velcro® to the underside of each of your solderless breadboards. When you want to use a board, just press the Velcro® on the board to the Velcro® on the square of wood, helping your breadboard stay put while you work on it. With this trick, you don't have to worry about the breadboards sliding around and pulling their wires out of their sockets.

Breadboard manufacturers make contact strips from a springy metal coated with a plating. The plating prevents the contacts from oxidizing, and the springiness of the metal allows you to use different diameter wires and component leads without seriously deforming the contacts. However, you can damage the contacts if you attempt to use wire larger than 20 gauge or components with very thick leads. If the wire is too thick to go into the hole, don't try to force it. Otherwise, you can loosen the fit of the contact, and your breadboard may not work the way you want it to.

 When you're not using it, keep your breadboard in a resealable sandwich bag. Why? To keep out the dust. Dirty contacts make for poor electrical connections. Although you can use a spray-on electrical cleaner to remove dust and other contaminants, you make things easier on yourself by keeping the breadboard clean in the first place.

All sizes, big and small

Solderless breadboards come in many sizes. Breadboards with 550 contact points accommodate designs with about three or four 14- or 16-pin integrated circuits, plus a small handful of resistors and capacitors.

For the most flexibility, get a double-width board, such as the one you see in Figure 11-3. This style accommodates at least 10 ICs and provides over 1,200 contact points. If you're into really elaborate design work, you can purchase extra large breadboards that contain 3,200 contacts or more.

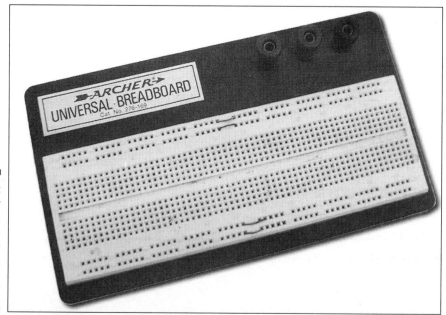

Figure 11-3:
For larger circuits, you can use bigger solderless breadboards.

Obviously, don't overdo it when buying a solderless breadboard. You don't need to buy one the size of Wyoming if you're making only a small circuit to power a light bulb. If you get into the middle of designing a circuit and find that you need a little more breadboard power, remember that some solderless breadboards have interlocking ridges so that you can put several together to make a larger breadboard.

Creating a Circuit with Your Solderless Breadboard

Essentially breadboarding consists of putting components onto the board with wires. But there's a right way to do things and a wrong way. This section gives you the lowdown on what type of wire to use, efficient breadboarding techniques, and the ins and outs of giving your board a neat, logical design.

Why you gotta get pre-stripped wires

You have to use solid (not stranded) insulated wires to connect components together on your breadboard. You should use 20- to 22-gauge wire. Thicker or thinner wire doesn't work well in a breadboard. Too thick and the wire won't go into the holes; too thin and the electrical contact will be poor.

As we suggest in Chapter 5, stay away from stranded wire. The individual strands can break off, lodging inside the metal contacts of the breadboard.

While you're buying your breadboard, purchase a set of pre-stripped wires. (Don't get cheap now; this is worth it.) Pre-stripped wires come in a variety of lengths and are already stripped (obviously) and bent, ready for you to use in breadboards. A set of pre-stripped wires costs from $5 to $7, but you can bet the price is well worth the time that you save. Otherwise, you need to buy a bunch of wire and painstakingly cut off about a ⅛ inch of the insulation on each end.

The assortments that you find in stores come with wires cut to different lengths. Table 11-1 gives you a rundown of the lengths and quantities from one popular assortment.

Table 11-1	Pre-Stripped Wire Lengths and Quantities
Length	*Quantity*
¾ inch	10
1 inch	20
1¼ inch	25
1½ inch	25
2 inch	10
2½ inch	10
3 inch	10
4 inch	5
6 inch	5

You may find that you need more wires of some lengths than you get in the assortment, so buy two. But the odds are that one day you will be missing one length of prestripped wire, and you'll have to strip a wire or two. You can actually cut your own lengths, and then use a wire stripper to do the job. It's best to use a stripper that has a dial adjustment. Set the dial for the gauge of wire that you're using, say 20 or 22. This setting prevents the stripper from nicking the wire; nicks weaken the wire. A weak wire can get stuck inside a breadboard hole, which can ruin your whole day.

To make your own breadboard wire, follow these steps:

1. **Cut the wire to length.**

2. **Strip off about ¼- to ⅛-inch of insulation from both ends.**

   While stripping the insulation, insert one end of the wire into the stripping tool and hold the other end with a pair of needle-nosed pliers. If you have an automatic wire-stripper/cutter tool (available at some hardware stores; check the electrical parts section), you can cut the wire and strip the insulation in one easy step.

3. **After stripping the insulation, use a pair of needle-nosed pliers to bend the exposed ends of the wire at a 90-degree angle, as you can see in Figure 11-4.**

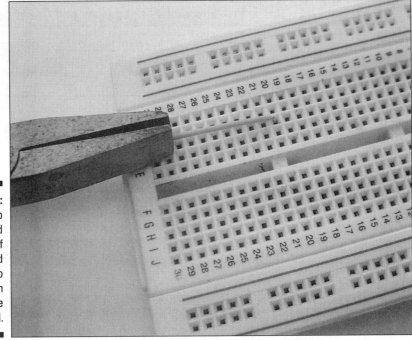

Figure 11-4:
You strip
and bend
the ends of
breadboard
wire to
insert them
into the
board.

Breadboarding techniques

Over the years, through trial and error, we've discovered some tips for using solderless breadboards. To save you the painful learning curve, here are some of our favorites:

✔ Use a chip inserter/extractor (most stores that sell electronics stuff sell these; they are about the shape of a ballpoint pen) to implant and remove ICs (integrated circuits). This nifty tool reduces the chances that you damage the IC while handling it. If you're working with CMOS chips, which are especially sensitive to static electricity, ground the inserter/extractor tool to eliminate stray static electricity. (If you need a refresher on integrated circuits, both the CMOS and TTL varieties, take a look at Chapter 4.)

✔ When you're using CMOS chips, build the rest of the circuit first. If you need to, use a dummy TTL IC to make sure that you wired everything properly. TTL chips aren't nearly as sensitive to static as the CMOS variety. Be sure to provide connections for the positive and negative power supply and that you connect all inputs (tie those inputs that you're not using to the positive or negative supply rail). When you're ready to test the circuit, remove the dummy chip and replace it with the CMOS IC.

✔ When inserting wire, use a pair of small needle-nosed pliers to plug the end of the wire into the contact hole. If the wire is too short, use the pliers to gently pull it out of the hole when you're done with the breadboard.

✔ Never expose a breadboard to heat, because you can permanently damage the plastic. ICs and other components that become very hot (because of a short circuit or excess current, for example) may melt the plastic underneath them. Touch all the components while you have the circuit under power to check for overheating.

✔ Solderless breadboards are designed for low-voltage DC experiments. They're not designed for, nor are they safe, carrying 117 VAC house current.

✔ You won't always be able to finish and test a circuit in one sitting. If you have to put your breadboard circuit aside for a while, put it out of the reach of children, animals, and the overly curious — you know the type, people who seem to always poke their fingers where they don't belong.

Neatness counts

You can easily build a bird's nest on your breadboard by routing connection wires carelessly. Neatness and tidiness are the keys to success when using a solderless breadboard. Messy wiring makes it harder to debug the circuit, and a tangle of wires greatly increases the chance of mistakes. Wires pull out when you don't want them to. Worse, the tangle of wires can cause the circuit to malfunction altogether. Chaos ensues.

Carefully plan and construct your breadboard circuits. This planning requires more time and patience on your part, but after you build a few projects, you find that the extra effort is well worth it. If you follow the advice in the next three sections, you'll greatly improve the neatness factor of the circuits that you build on your solderless breadboard.

Avoiding the crowd

Give yourself enough room to move around. If your circuit uses integrated circuits, start with those and provide at least three to five columns of holes between each IC. Go for ten empty columns between each IC if you can. Then add the other components.

If the breadboard is too small to accommodate all the parts, switch to a bigger one. Or, if you don't have a bigger breadboard, purchase a second one and string them together.

Don't worry about urban sprawl on your breadboards. You do better to place the components a little farther apart than to jam them too close together. Keeping a lot of distance between ICs and components also helps you to tweak and refine the circuit. You can more easily add parts without disturbing the existing ones.

Logical layouts

The row/column arrangement of breadboards invites you to create a rather haphazard layout of a circuit. One approach is to put the primary component, such as a 555 timer IC or a microcontroller, in the middle of the board and work your way out.

Here are more layout ideas:

- ✔ If you can, place the components in a way that reduces the number of *jumper wires* (a jumper wire can be any old wire; it's the fact that you use it to jump from one connection to another that gives it its name). The more wiring that you have to insert, the more crowded the board becomes and wires can come loose.

- ✔ Don't be afraid to clip some extra length off the leads of components. Resistors, capacitors, and diodes don't cost much. Cutting off excess leads lets you lay components flat against the board, which helps them stay in place (see Figure 11-5). Save these parts for use on another circuit, or just toss them, if the leads are too short.

- ✔ Use a small pair of pliers to bend the leads and wires to a 90-degree angle. Keep the wires as close to the board as possible. This positioning helps prevent the wires from pulling loose while you're working on the circuit.

Establishing common connection points

The two most common connections in most circuits are power and ground. Solderless breadboards provide a convenient point for these connections in the long rows along the top or bottom.

If a circuit requires other common connection points, and you don't have enough points in one column of holes (breadboards usually have five holes per column), use longer pieces of wire to bring the connection out to another part of the board where you have more space. You can make the common connection point one or two columns between a couple of integrated circuits, for instance.

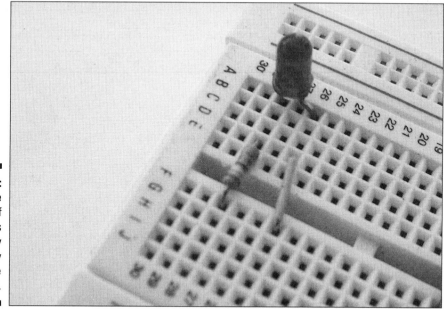

Figure 11-5:
Clip the length of components so they fit neatly on the breadboard.

Making the Move from Your Circuit to a Solder Breadboard

So you've perfected the world's greatest circuit, and you want to make it permanent. The next step after the solderless breadboard is the solder breadboard, also called a solder board, an experimenter's PC board, or a universal solder board. The solder breadboard allows you to take any design that you create on a solderless breadboard and make it permanent. You can transfer your design to a solder board easily because the solder breadboard has the exact same layout as the solderless breadboard.

To transfer your design, you simply pick the parts off your solderless breadboard, insert them in the solder breadboard, and solder them into place in the corresponding spots. Use wires like you did in the original solderless breadboard to connect components that aren't electrically connected by the metal strips of the circuit board. (If you're new to soldering, and what's involved, be sure to read Chapter 8.)

If you design a really small circuit, you can use just one half of a solder breadboard. Before transferring the components, cut the solder breadboard with a hacksaw. Try not to breath in the dust produced by the saw. Clean the portion of the board that you want to use and solder away. See Chapter 12 for more information on cutting and cleaning circuit boards.

My breadboard circuit doesn't work right!

By now, you realize that electronics is a screwy science. Sometimes things work, and sometimes they don't. But you're not dealing with the fickle finger of fate; electronics misbehave for very real, understandable (and mostly fixable!) reasons. As you work with breadboards, you may encounter the fairly common problem of stray capacitance. This condition occurs when components and wires produce unwanted capacitance (stored charge) in a circuit. This situation can happen when a bunch of wires criss-cross, for example.

The reason this occurs is a bit complex, but it has to do with the strips of metal used inside the breadboard, and also the longer lead lengths of the components. All circuits have an inherent capacitance. It can't be avoided. When there are lots of wires going every which way, the capacitance can unexpectedly increase—this is stray capacitance. At a certain point (and it differs from one circuit to the next) the changes in capacitance can cause the circuit to misbehave.

Although most circuits that you test with a solderless breadboard work well enough, some projects behave erratically or unpredictably if you don't build them by using a printed circuit board and solder. You really need to be aware of this fact if you're working with RF (radio frequency) circuits, such as radio receivers and transmitters, digital circuits that use signals that change at a very fast rate (on the order of a couple of million Hertz), and more sensitive timing circuits that rely on exact component values. Solderless breadboards have a tendency to change the characteristics of some components, most notably capacitors and inductors; these variations can change the way a circuit behaves.

If you're building a radio or other circuit that stray capacitance can affect, you may have to forego the step of first building the circuit on a solderless breadboard. You may have better luck with the performance of the circuit by going straight to a solder breadboard or another type of soldering board, which we describe in the section "Prototyping with Pre-Drilled Perf Boards."

Leave space at the corners of the board so that you can drill mounting holes. You use these holes to secure the board inside whatever enclosure your project provides (such as the chassis of a robot). Alternatively, you can secure the board to a frame or within an enclosure by using double-sided foam tape. The tape cushions the board and prevents breakage, and the thickness of the foam prevents the underside of the board from actually touching the chassis.

Solder breadboards have one main disadvantage: They don't use space very efficiently. Unless you cram the components onto the board, the breadboard limits you to building circuits with only two or four integrated circuits and a handful of discrete components. In time, you can figure out how to conserve space and make good use of the real estate on a solder board.

Prototyping with Pre-Drilled Perf Boards

Solder breadboards aren't the only kind of general-purpose circuit board that you can use for your projects. You have another option in a pre-drilled perf board with copper traces for wiring. These boards go by many names, such as a grid board or a universal, general-purpose, or prototyping PC board. Perf boards come in a variety of sizes and styles. Figure 11-6 shows a few perf board styles. All styles are designed for you to use with ICs and other modern-day electronics components, which means the holes are spaced 0.100 inch apart.

You may find perf boards handy when you want a soldered circuit but don't want to go through the much more laborious process of making your own circuit board from scratch. One of the main ways that you use perf boards is to construct circuits by using wire wrapping (see the section "Getting Wrapped Up in Wire Wrapping," later in this chapter for more about this). If the board has pads and traces on it, and most do, you can solder components directly onto it. But you can use a perf board without any copper pads and traces when you use the wire wrapping method.

Figure 11-6:
A few sample perf boards, ready for you to clean them if necessary and add electronics components.

You can choose the style of grid board based on the type of circuit that you're building. Some grid layouts suit certain applications better than others. Personally, we prefer the plain universal PC board with interleaved busses, as we think they are easier to use. (You don't ride the buss on a circuit board, you solder things to it. A *buss* in the electronics world is just a common connection point.) You tie components together on the universal PC board, using three- (or more) point contacts.

A buss runs throughout the circuit board so that you can easily attach components to it. Many perf boards have at least two busses, one for power and one for ground. The busses run up and down the board, as you can see in Figure 11-7. This layout works ideally for circuits that use many integrated circuits. Alternating the busses for the power supply and ground also helps to reduce undesirable inductive and capacitive effects.

You use the perf board just as you use a solder breadboard. After cleaning the board so that the copper pads and traces are bright and shiny, plug the components into the board and solder them into place. Use insulated wire to connect components that aren't adjacent to one another.

Figure 11-7:
Several busses run up and down this perf board.

Making circuit boards with plug-'n'-play ICs

You may want to consider this idea when you build circuit boards that include integrated circuits: Instead of soldering the IC directly onto the board, use an IC socket. You solder the socket onto the board, and then, when you're done soldering, you plug in the IC and hit the switch.

IC sockets come in different shapes and sizes, to match the integrated circuits they're meant to work with. For example, if you have a 16-pin integrated circuit, choose a 16-pin socket.

Here are some good reasons for using sockets:

✔ Soldering a circuit board can generate static. You can avoid ruining CMOS or other static-sensitive integrated circuits by soldering to the socket rather than the actual IC.

✔ ICs are often one of the first things to go bad when you're experimenting with electronics. The ability to pull out a bad chip and replace it with a working one makes troubleshooting a whole lot easier.

✔ You can share an expensive part, such as a microcontroller, among several circuits. Just pull the part out of one socket and plug it into another.

Sockets are available in all sizes to match the different pin arrangements of integrated circuits. They don't cost much — just a couple of pennies for each socket.

Getting Wrapped Up in Wire Wrapping

Wire wrapping is a point-to-point wiring system that uses a special tool and extra-fine 28- or 30-gauge wrapping wire. When you do it properly, wire-wrapped circuits are as sturdy as soldered circuits. And you have the added benefit of being able to make modifications and corrections without the hassle of desoldering and resoldering.

You have to limit wire wrapping to projects that use only low-voltage DC. It's not for anything that requires a lot of current, because the wire you use isn't large enough to carry much current.

To wire wrap, you need

✔ **Perf board:** You attach the components to this board. You can use a bare (no copper) board or one that has component pads for soldering. We personally prefer the padded board.

✔ **Wire-wrap sockets for ICs and other parts:** These sockets have extra-long metal posts. You wrap the wire around these posts.

✔ **Tie posts:** These posts serve as common connection points for attaching components together.

- **Wire-wrap wire:** The wire comes pre-cut or in spools. We prefer pre-cut wire, but try both before you form an opinion.

- **Wire-wrapping tool:** You only have to use this specific tool to wrap wire around a post and remove it. The tool also includes an insulation stripper; use this, not a regular wire stripper, to remove the insulation from wire-wrap wire.

Though you can wire wrap directly to resistor, capacitor, diode, and other component leads, most people prefer using wire-wrap sockets. The reason? Most components have round leads. A wire-wrap socket has square posts. The square shape helps to bite into the wire, keeping things in place. If you wrap directly to component leads, you may want to tack on a little bit of solder to keep the wire in place.

Here's the basic process for wire wrapping:

1. **Insert a socket into the perf board.**

    If the board has solder pads, touch a little solder between one of the pads and the post sticking through it. This dab of solder keeps the socket from coming out.

2. **Repeat Step 1 to insert all the other sockets that you may need.**

3. **Use the wire-wrap tool to connect components together.**

4. **Plug the ICs and other components into their sockets.**

A big advantage of wire wrapping is that you can make changes relatively easily. Simply unwrap the wire and re-route it to another post. If the wire gets cruddy, just replace it with a new one.

There's more to wire wrapping than we can cover here. If it sounds like a method you think would be useful to you, do an Internet search on "wire wrapping techniques" (include the quotes for more specific results). You can find numerous Web sites that help you become an expert wire wrapper-upper, such as http://www.me.umn.edu/courses/me2011/robot/wrap/wrap.html or http://www.okindustries.com/products/4.1.1.1.htm#The.

Chapter 12

Building Your Own Printed Circuit Boards

*Y*ou're well on your way to being an electronics guru when you make your first printed circuit board. Forget all the wires criss-crossing from one place to the next on a breadboard or components stretched out over a pre-made soldering board. You know that you're playing with the big kids when you hold up your very own custom-made circuit board, and say "I made this!"

To begin with, in this chapter, you discover just what makes up a circuit board. The electronic construction technique of choice for many is the printed circuit board, or PCB. There are a number of ways to build a custom printed circuit board for your project. This chapter details several methods, including direct etch, photo transfer, and laser film transfer. We also give you tips about how to use a computer to speed up your work.

Anatomy of a Circuit Board

Before we get into how to make a printed circuit board, we take a closer look at what makes up the typical PCB:

✔ You make a printed circuit board by gluing a very thin sheet of copper over a plastic, epoxy, or phenolic base. This copper sheet is called *cladding* and represents the foil side of the board. We show an example of this bit of copper in Figure 12-1. It looks pretty boring because here it's just a blank canvas. It can become almost anything at this point.

✔ To make the final board, you etch away specific portions of the copper, leaving just the printed circuit design. We talk about the exact methods of laying out the circuit and etching the board in more detail in the section "Showing You My Etchings: Etching the Circuit Board," later in this chapter.

✔ You produce the circuit with pads and traces:

  • **Pads:** These contact points for components are generally round or rectangular in shape (these are called "donuts" in electronics parlance). After you etch the circuit board, you drill a hole in the center of each pad. You mount the components on the top of the board with their leads poking through the holes. You then solder each component lead to the board at the board's pads.

  • **Traces:** These wires of the circuit board run between the pads to electrically connect the components together. See Figure 12-2 for an example of traces.

✔ Printed circuit boards can be either single- or double-sided:

  • **Single-sided boards** are copper clad on only one of their sides. You mount the components on the other side.

  • **Double-sided boards** are copper clad on both sides; you often use these boards when you're working with a very complex circuit. It's hard to make your own double-sided PCBs, but you can design them and have them made for you. We tell you more about PCB manufacturers in the last part of this chapter.

More advanced circuit boards have multiple layers. An insulating covering keeps the layers from shorting out. Multi-layer circuit boards are way beyond what most people can make on their own, so we just mention them here and go on our merry way.

Figure 12-1:
A printed
circuit
board is
made
of copper
and an
insulating
backing.

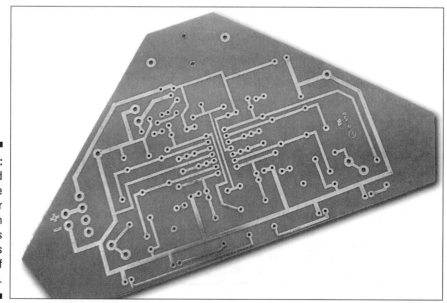

Figure 12-2:
Finished
PCBs use
pads for
soldering on
components
and traces
in place of
wiring.

How the Copper Gets onto the Circuit

There's more than one way to skin a cat, and there's more than one approach to building a circuit board. To begin with, you can use a number of methods to form the pads and traces that ultimately turn a sheet of copper into a bona fide PCB.

Here's how most commercial manufacturers make circuit boards. While you may not create your own circuit boards in the same way, it's helpful to know how the process is typically done

1. **First, they coat the copper with a light-sensitive chemical layer called the *sensitizer*, also known as *resist* or *photoresist*.**

2. **Next they place an exact-size film negative of the circuit board layout drawing over the copper clad and expose it.**

   Just as in processing a photograph, they expose the negative by bathing the board in light, in this case, strong ultraviolet light. The light passes through the negative and strikes the sensitized copper underneath.

3. **After exposing the board, they dip it in a *resist developer* (this is messy stuff but a necessary part of the process).**

   What comes out from the developer is a copper-clad board where the portions of the copper that weren't exposed to light have turned black or a dark gray. As a point of reference, they often refer to this process as the *positive method;* the *negative method* produce blacks or grays on those areas that are exposed to light.

4. **As a final step, they dip the board in etchant solution.**

   The *etchant* is a strong acid-like liquid that eats away at the copper. The black/gray areas resist the action of the etchant, forming the circuit pattern on the board. (You can find more information about etchant and etching in the section "Showing You My Etchings: Etching the Circuit Board," a little later in this chapter.)

You can duplicate this photographic method in your own shop, although the process takes a lot of time and can cost you a boatload. You can purchase all the chemicals that you need from specialty electronics supply outfits; check out the Appendix at the end of this book for a list of sources. However, the rest of this chapter concentrates on simpler methods that are probably more up your alley.

Ready, Set: Preparing to Build Your Board

As with most things that you do in electronics, a little planning and neatness count for a lot. Before jumping into actually building a circuit board, you should know how to choose and prepare materials.

Choosing the right copper clad

Materials that you use to make printed circuit boards come in various forms. You can buy either single- or double-sided copper clad. Unless you're making a double-sided board, stick with single-sided clad. You waste etchant if you make single-sided boards with double-sided clad, and besides, the double-sided stuff costs more.

The thickness of cladding also varies. A common thickness is 1 ounce per square foot. That's about 35 micrometers, or a little less than half the thickness of a human hair. You don't need to worry about the exact thickness of the clad with most hobby circuits, except when you're using high voltages or currents. For most of the circuits that you build, you can opt for just about any cladding thickness that you want. But remember that if you get it too thick, you just use up the etchant faster.

The thickness of the board material that you're cladding also varies. A G-10 grade board, with the standard thickness of 0.062 inch, is made of epoxy resin and is perfect for the job. Paper-based phenolic and epoxy resin boards, usually referred to as FR-2 or FR-3, are flame resistant and cost more. The temperature and flame-resistant FR-4 and FR-5 grade boards also take a bit more cash than standard boards. For most hobby uses, just pick the stuff that takes the least amount of money out of your wallet.

Cutting and cleaning

The board that you use to make your PCB should be only as large as the circuit layout, no larger. Etching a 5 x 5-inch board that contains only 2 x 3 inches of circuit space wastes etchant solution and money.

If the board is too large, cut it with a saw. First draw or scribe a line in the copper to ensure a straight cut. You can then smooth the edges of the board with a file or fine-grit sandpaper. Wear eye protection and a face mask to keep the dust from the circuit board out of your eyes, nose, and mouth.

Going surplus

You can sometimes find surplus copper clad that's a cut-off or a remnant from a larger piece used in the electronics industry. This stuff works perfectly well, though it may be grimy or dirty. You may find such surplus scraps if you have an electronics company nearby. Most of those folks just throw this stuff away. Or, you can buy the clad surplus from surplus suppliers. Check out the Appendix for leads on some good mail-order surplus suppliers for electronics.

You probably have to cut your PCB pieces down to size if you buy them as surplus scraps. Use a sharp metal shear to cut the board to size, rather than a saw. The reason? Most PCBs use an epoxy and fiberglass base; when you cut the base, fine dust particles from the fiberglass float in the air. These fiberglass particles can irritate your eyes and nose. No matter what tool you use, if you must cut circuit board pieces to size, wear eye protection and a respirator mask.

No matter what circuit-making process you use, be sure that you completely clean the copper before you apply any chemicals to it. Dirt and grease foul up the etching process. Scrub the copper, using a non-metallic scouring pad (such as the one in Figure 12-3), and a household cleansing powder. Scrub for a minute or two to remove all grease, dirt, and oxidation.

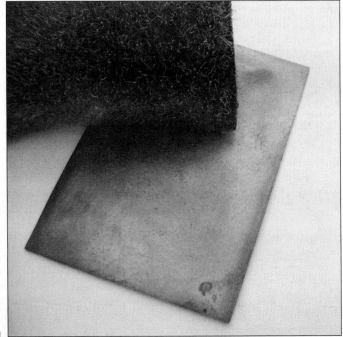

Figure 12-3:
The copper clad must be spotless before you make a circuit on it.

When you have the copper really clean, the water should form a smooth sheet on the surface, not bead up like water on a newly-waxed car.

Creating a PCB Photographically

The easiest way to make a printed circuit board is to use a pre-printed template, such as those templates that you find in some electronics magazines and books. You can photograph these templates and copy them to acetate film.

In the following sections, we explain some of the considerations and procedures that you use to turn existing artwork into a circuit board. Here is an overview of the process you use to make a printed circuit board using the photographic method (it is quite similar to the process the manufacturing pros follow, described above in " How the Copper Gets onto the Circuit":

 ✔ Prepare a mask of the circuit layout and transfer it onto clear transparent film. This is done using a variety of methods, as described in the sections that follow.

 ✔ Use the mask to expose a sheet of sensitized copper to strong ultraviolet light.

 ✔ Dip the sensitized, exposed sheet into a developer chemical. This produces a pattern (called a resist pattern) of the circuit board layout. The pattern matches what was exposed through the mask.

 ✔ Submerge the copper sheet into a tray of etchant. The etchant effectively washes off the copper everywhere except under the resist pattern.

Making the mask

Using the photographic method, once you make a photographic positive or negative of the original artwork you then use this photographic copy as a mask to expose an image of the PCB onto a piece of photosensitized copper (you learn more about this in the section "Positively or Negatively Sensitized?").

So your first step is to decide if you want to make a positive photographic copy of the artwork or a negative copy. You make the copy on a thick film, like the film in a 35mm camera (only larger and without the holes!):

 ✔ A **negative** inverts the polarity of the image; black areas become clear, and white areas become black.

 ✔ A **positive** retains the polarity of the image; black areas remain black, and white areas become clear.

The art of layout drawings and schematics

Although both layout drawings and schematics are representations of a circuit, there are differences. A schematic shows the components of the circuit and what connections run among them. A layout drawing essentially shows the traces you use to make the connections that are called for on the schematic. While all the connections are shown in a schematic as lines, the lines and components don't accurately reflect the physical layout on a final board. In a layout drawing, these connections are shown as they will actually appear on the board.

Have a local printer produce a photo negative or positive of the original artwork that you find in a magazine or book. It costs about $10 or so, but it can really help when you have to make multiple copies of a circuit board.

As an alternative, you can try making a positive mask with a sheet of transparency film. The local copy shop can probably do a better job (they get deeper, darker blacks) than your printer at home. If you photocopy the original onto transparency film, be sure to avoid any sizing errors. Many copiers automatically apply a 1- to 2-percent enlargement, and this size change can slightly alter the dimensions of the solder pad and hole spacings. Be sure to get an exact 1:1 copy.

If you take the artwork to a print shop, they can usually make a negative for less money than a positive because fewer steps are involved. When you make a photocopy onto transparency film, you always get a positive image. Whether you use negative or positive art, be sure to select the proper sensitizer and developer to match. More about this stuff in the section "Positively or Negatively Sensitized," immediately following.

Creating the film has two most important aspects: The black areas need to be completely black and filled in, and any thin lines in the circuit can't break up. Closely inspect the film under a bright, even light to look for imperfections. Sometimes, you can just fill in the missing black areas with a Sharpie® or similar marking pen.

Positively or negatively sensitized

We made a bit of a fuss about making a positive or negative in the previous section. Here's where it matters: You have to be sure to use the right kind of sensitized copper clad or sensitizer (also called photoresist) spray to match the artwork that you're using. The sensitizer, whether the manufacturer already applied it to the board, or you spray it on, is what makes the copper "pick up" the pattern from the artwork.

Most circuit board layouts in magazines and books are designed as film positives where the black portions represent areas that contain resist, and the etchant washes away the white areas.

If you make a film positive to use as the mask, you must use a positive-acting sensitizer. When you expose the board to light, the dark part of the artwork stops anything underneath it from being exposed. When you dip the board into the developer, the unexposed portions take on a darkish coating of the resist. When etched, the part of the board under this resist remains. The rest of the board is removed, because there is nothing protecting the copper. Conversely, a negative-acting sensitizer turns the clear areas of the artwork into resist. You end up with a board that matches the mask, as long as the artwork and sensitizer/developer match, positive-wise or negative-wise.

You can buy sensitizer solution that you spray onto ordinary copper clad boards. Or, you can buy boards that come already sensitized, which is a lot easier. You can tell they're sensitized because they come in black plastic baggies. Boards that somebody else sensitizes cost you more, but the time savings are worth it.

You also need a developer that's compatible with the type of sensitizer that you're using. When you buy sensitizer spray, the manufacturer usually packages the correct developer with it. Be sure to not mix a positive sensitizer with a negative developer, or your board pattern won't come out correctly.

Mirror, mirror on the PCB

Okay, one more thing to be careful about: Be sure that you transfer the layout to the board with the proper orientation. Reversing the layout makes a mirror-image of the board, leaving you with a pretty useless circuit board. If you reverse the layout, the connections for any integrated circuits are backwards; at best, the circuit just doesn't work, and at worst, it can burn out components. Spend some time thinking about how the layout transfers to the copper clad, and be sure that you don't reverse the layout.

Preparing the PCB for etching

After you make a mask and sensitize the copper clad (or purchase presensitized copper), you're ready to actually make the printed circuit board.

Here's what you do to make the board:

1. **In a darkened room (remember, this is a photographic process so you need to work in the dark!), spray a coating of sensitizer onto a clean copper-clad board.**

2. **Place the film over the sensitized board and insert both in a suitable exposure holder, like the picture frame in Figure 12-4.**

   To be really fancy, you can get photographic holders at a photography shop, but picture frames work just as well.

Figure 12-4:
Use a picture frame to hold your circuit board and artwork during exposure.

3. **Note the orientation of the film and correct it, if necessary.**

   Make sure that you haven't reversed the film, or you may etch the board with a mirror-image of the circuit layout. Checking the film's position requires some thought on your part, so don't rush through it.

4. **Place the film negative so that the emulsion (the dull side) faces the copper clad. This step helps produce a sharper image on the board.**

Note, however, that you may need to reverse the film (emulsion side facing out) if the original artwork that you're copying from a printed magazine or book is reversed. Some magazines print circuit artwork in reverse (called left-reading), to make it easier for them to transfer the image using the dry toner transfer method, described in the section, "Creating a PCB By Using the Transfer Film Method," later in the chapter.

Let there be light: Exposing and developing the board

You can expose the sensitized board in a number of ways. If the sun is out and about, you can expose the board to its ultraviolet rays. Exposure time varies between just a few minutes to over 15 minutes. The many makes and brands of photoresist require different exposure times, so check the instruction sheet that came with the photoresist that you're using.

You can reduce exposure time by using an ultraviolet tanning lamp. Place the lamp a foot or two away from the board so that the light falls evenly on the board. Don't put the lamp so close that the edges of the board are in shadow.

Remember that UV tanning lamps give off ultraviolet rays, so don't expose yourself to these rays for long periods, unless you actually enjoy the stinging sensation of sunburn! Also, wear tanning bed goggles that keep the UV rays from entering your eyes to avoid permanent damage to your retinas. Better yet, don't use an ultraviolet lamp. Wait for a sunny day and expose the board in the great outdoors.

Which end is up? (Or down or left or right?)

If you want to make sure that you orient the film properly, try this approach. Look at the component layout diagram that you can usually find printed with the circuit layout. This diagram shows where you should place each component on the board. The shaded portions of the component layout diagram show where the pads and traces are located on the underside of the board.

Now look at the film and orient it to match the shading of the component layout diagram. Most boards aren't symmetrical, so you can easily tell the left side from the right side. Place the sensitized board into its exposure holder. Finally, flip the film over (left to right) and place it over the board.

After you expose the board, you have to develop it in a suitable developing solution. Mix the developer according to the instructions the manufacturer included with the package. After you mix it, pour the developer into a shallow plastic tray — the kind you find at a photography store is perfect. Place the exposed board into the developer liquid. Developing times vary from one manufacturer to another. Follow the solution manufacturer's recommended times for developing.

After developing, you're ready to etch the board, which we detail in the section titled "Showing You My Etchings: Etching the Circuit Board," later in this chapter.

Creating a PCB by Using the Transfer Film Method

The photographic method that we describe in the previous section requires that you use sensitized copper clad, a negative (or a positive), and exposure to ultraviolet light. All in all, this process means a lot of fuss if you're making a slew of boards.

If you plan on making just one or two circuit boards from pre-printed artwork, you may want to consider the transfer film method. The transfer film method involves nothing more than a sheet of clear acetate — the type used for overhead transparencies — and a plain paper copier or laser printer.

When you copy the artwork onto the transparency, it fuses black toner onto the acetate sheet. You transfer the toner from the transparency to the copper clad using the heat from a clothes iron. The toner provides an effective resist to the etchant solution that you use to dissolve unwanted copper from the circuit board. Although the method may seem anything but high-tech, it works quite well.

Although overhead transparency acetate works as a transfer film medium, you get better overall results if you use transfer film specifically designed for the job. Check out the Appendix at the back of this book for a list of several online resources that sell transfer film designed for making printed circuit boards.

You need a plain paper copier or printer in top-notch working condition, otherwise the image that the transfer film records turns out gray and grainy. If you don't have such a copier nearby, take your printed circuit board artwork to your local copy center. Many neighborhood copy shops use state-of-the-art high speed plain paper copiers that produce jet-black images on clear acetate film. We think the few extra pennies that you spend are well worth it.

By the way, some copiers require that you use acetate sheets with a leading white strip. The copier uses this strip to sense when a piece of paper feeds through critical parts of the machine. Without the strip, the acetate may jam. Before buying a box of transparency film for your copier, verify what kind of film your copier works with. Most stationery stores don't let you return opened boxes of copier supplies.

Flip-flop, flop-flip

As with the photographic method, which you can read about in the section "Creating a PCB Photographically," earlier in this chapter, you need to make sure that you have the circuit board image oriented the right way. The artwork must be right-reading, meaning it shouldn't be reversed left-to-right. You have to again reverse original artwork that you get from a book or magazine that the publishers already reversed once (left-reading, or mirror image).

You can verify that you have the correct orientation by

- ✔ **Using a plain paper copier that can reverse the image.** Some of the top copier models have this feature.

- ✔ **Making one copy of the artwork onto the transfer film and then turning the film over and using it to make a second copy.** Place a piece of white paper behind the transfer film for a clean background. This method isn't ideal, but it works in a pinch.

You can tell if the image is right- or left-reading by looking at the text that accompanies the PCB layout. The image is right-reading if the text appears normally. If the text appears backwards, the image is left-reading.

Getting a good image

After you copy the artwork to the transfer film, look carefully for washed-out areas or areas where toner hasn't completely adhered. Gently touch it with a soft tissue to make sure the toner is firmly bonded to the sheet. Some copiers don't fuse toner well to overhead transparency film. If you find out that your copier can't cut the transparency-film mustard, try a different copier which may have a higher heat setting.

If the image looks good, protect the transparency by covering it with a blank sheet of paper. If you accidentally scrape toner off the film, you can cause a void in the printed circuit.

Transferring the layout to copper clad

After going through the steps in the previous two sections, you're ready to transfer the toner from the transparency to the copper clad of the printed circuit board.

Follow these steps to prepare the board:

1. **First, clean the board thoroughly by using a household cleanser, such as Ajax or Comet, and a sponge. After cleaning be sure to *not* touch the copper. Handle the board by the edges only.**

   Be absolutely sure there is no oil or dirt on the board. The copper should be bright and shiny.

2. **Place the PCB copper-side up on an ironing board.**

3. **Cut the transfer film to size so that it's not larger than the board.**

4. **Carefully place the transfer film over the board so that the toner side faces the copper clad.**

5. **Secure the film into place with some strips of masking tape, but don't tape over toner areas.**

6. **Place a small piece of cheese cloth or several layers of paper towel over the PCB and transfer film. (Cheese cloth is a thin cotton material you can get in the housewares aisle of your nearby department or grocery store. You can also use plain white paper towels (no pretty pastel flowers printed on them, please).**

Now, it's time to transfer the artwork to the board. If you're using transfer sheets specifically designed for making PCBs, refer to the instructions that come with the transfer sheet.

The following steps provide general recommendations for the best way to transfer if you're using ordinary transparency film:

1. **Set your clothes iron to cotton-linen or medium-high heat.**

2. **Let the iron warm up and then apply the iron to the board. Move the iron back and forth in slow, even strokes, as if you're ironing your very best shirt.**

   Be careful to keep the cheese cloth or towel flat to avoid wrinkles.

3. **Apply steady and firm pressure for 15 to 20 seconds (see Figure 12-5).**

4. **Wait 10 to 15 seconds, then gently lift the cloth.**

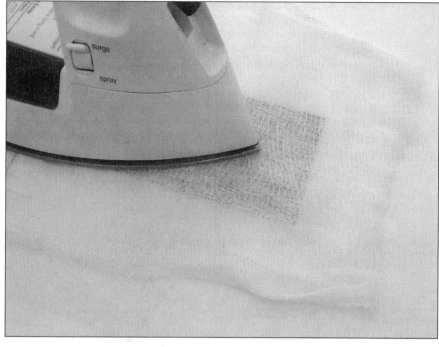

Figure 12-5:
Use a clothes iron to transfer the toner from the transparency to the copper clad.

5. **Carefully peel back a small corner of the film to see if the toner has transferred to the copper clad.**

6. **If the toner hasn't completely transferred to the copper, replace the film and apply the iron for another 15 to 20 seconds.**

You have to experiment with the exact time to achieve proper toner transfer. It may take as little as 15 seconds or as long as a minute. Don't try to speed things up by increasing the temperature setting of the iron. This increase makes the transparency film wilt, causing a distorted transfer of the circuit layout.

Be sure to QC (Quality Control) your work!

After you've transferred the toner, wait for the film to cool. If it's hot to the touch, leave it for a minute or two. Then gently remove the spent film and discard it; you can't use it again.

Carefully inspect the pattern on the copper clad. The toner should be well adhered with few, if any, voids. Use a magnifying glass (3x or 4x) to get a close-up view. If you see voids or skips, fill them in with a fine-tipped resist pen, which you can find at most electronics stores, or with a Sharpie®. The pen uses a jet-black ink that resists the action of the etching solution.

You may find that you need to press the toner into the copper clad using a small wooden or rubber roller. The pressure of the roller helps fuse the toner to the copper. We have had good success with a wooden wallpaper seam roller, such as the one in Figure 12-6. Be careful to burnish the toner onto the copper with a rolling action, not a scraping action, or you may scratch off the circuit layout.

Figure 12-6:
Use a wooden seam roller to burnish the toner into the copper clad.

You now have the board ready for etching, which you can find out about in the section "Showing You My Etchings: Etching the Circuit Board," later in this chapter.

Choosing a Method for Making Your Own Circuit Layouts

Can't find any artwork on which you want to base your circuit board? You're probably glad to hear that you can also make a printed circuit board completely from scratch.

You can try two popular methods for making a board from scratch:

✔ **The direct-etch method:** With this method, you apply the resist directly to the copper clad. Dry transfer direct-etch kits, such as those made by Datak, provide pads for ICs and other components, as well as spools of thin black tape for the traces. Refer to the Appendix for Web sites that sell dry transfer kits. You apply the resist just like you make signs, using dry transfer lettering. The direct-etch method is practical for designs where you want to make just one board.

✔ **Draw the layout:** You can use a computer and plotter, ink pen, or other method to draw your layout and then use that drawing as a master to make one or more PCBs. The *master* is essentially a film negative or positive that you use to expose the surface of the sensitized copper clad. You can buy pre-sensitized boards or apply the sensitizer to a standard board with a brush or spray can. Exposure typically requires a strong short-wave ultraviolet light source, such as a tanning lamp.

Showing You My Etchings: Etching the Circuit Board

Creating the resist pattern on a new sheet of printed circuit board material, as we describe in the section "Making Your Own Circuit Layouts," earlier in this chapter, really only gets your board one third of the way done. For the next step, you need to etch that board to remove the unwanted copper. The copper that remains forms the printed circuit that makes your project work.

You use something called etchant to etch your board. *Etchant* is a caustic (meaning it can burn you) chemical that dissolves copper. It's not like some acid that a monster in a B-movie oozes out dissolving everything in its path; etchant doesn't fizzle away the copper on contact. The etching process actually takes several minutes. The copper that the resist pattern doesn't protect dissolves away first. The etchant finishes its job when it gets rid of all the copper in the exposed, resist-free areas.

(The final third of the PCB-making process involves drilling the holes for the components, which we cover in the section "Final Prep and Drilling," later in this chapter.)

First step: Inspecting the board

Think of etching as an unforgiving process. In the steps leading up to this process, you can modify or redo your work, to a certain extent. But when you reach the etching stage, you're making a commitment: After you etch, if you have an error in your circuit, you probably have to chuck the whole thing and start again.

That's why you really, really need to inspect the board for errors in layout, missing traces and pads, skips in the resist pattern, and other gremlins that produce a poor result before you actually etch the board:

- ✔ If you created the resist pattern from artwork appearing in a book or magazine, compare your board with the printed layout. Follow the traces from pad to pad and note any discrepancies. If you find any, you'll have to redo the artwork or fix any problems before you begin etching.

- ✔ If you created the resist pattern from your own design, or by using the direct-etch method, carefully review your work and compare it against a schematic or paper drawing. Be sure that pads and traces aren't too close together. At a minimum, all pads and traces should be ½nd of an inch apart, but more is always better here.

Repairing a board after a shoddy etching job — if you can do it at all — is time-consuming and frustrating.

Cleaning the board — carefully, please!

After you inspect the board, wet a cotton ball with isopropyl alcohol and gently clean the exposed parts. Don't apply much alcohol because some types of resist may melt or distort when exposed to alcohol. Also, let the alcohol dry completely before immersing the board in the etchant fluid.

Use isopropyl alcohol with a minimal water content. General purpose isopropyl alcohol that you buy at the drug store can have 30- to 40-percent water content. The more water mixed in with the alcohol, the more chance you have that the water will damage the resist. Look for so-called technical grade isopropyl alcohol, available at chemical supply outlets and school lab suppliers.

Kvetching about etching

Etching can be dangerous — not only to your health, but also to your wardrobe. Most circuit board etchants, whether in liquid or powder form, are toxic and highly caustic. Never allow the etchant chemical to come into contact with your skin or your clothes. If you do get some etchant on your fingers or hands, wash it off immediately.

Types of etchant chemicals

Perhaps the most common etchant chemical for making printed circuit boards is ferric chloride. You can buy it in both liquid and concentrated dry (or paste) form. The liquid often comes already diluted and ready for use. But you will have to dilute dry powder or paste in water. We recommend the concentrated form because you can mix it with hot water. The hotter the water, the faster the etching process. (But never exceed 135 degrees Fahrenheit or else the chemical may bubble and get all riled up, and splash over you and everything near you.)

Ammonium persulfate is gaining in popularity as a PCB etchant. It is available in both liquid and crystal form. Folks in the know about chemicals and safety generally consider ammonium persulfate safer than ferric chloride, though both chemicals are toxic. Handle both with extreme care.

Because etchant stains skin and clothing, avoid wearing your best party clothes when etching. Instead, wear a smock, your least favorite pair of pants, and old shoes. Also, wear eye protection to prevent the etchant from injuring your peepers if it splashes onto your face.

Wear gloves to protect your hands against burns and stains. Choose gloves that let you work almost as well as if you didn't have gloves on at all. (So don't use those old gardening gloves for this kind of work.) Disposable plastic or latex gloves do a good job.

Prolonged exposure to etching solution fumes can seriously injure you, so be sure to etch your circuit boards only in a well-ventilated area. All etchant solutions give off fumes, which can do serious harm to the mucous membranes in your nose and throat. You don't necessarily notice the effect right away. You may etch one or two boards and not be aware of the fumes. But an hour or two later, you feel an intense burning in your nose or throat that can last up to several days.

Store unused etchant solution in a dark-colored plastic bottle designed for photographic chemicals and keep the bottle in a dry, dark, cool place. Clearly label the bottle with its contents and keep it away from children.

Mixing the etchant

If the previous section didn't scare you away from ever touching etchant, even with a ten foot pole, check out this section for the mad scientist portion of the process — mixing the etchant.

Talking of trays and tongs

No matter what type of etching solution you use, always pour it into plastic containers. Avoid all metals because the etchant reacts with them. Be sure that the stopper or cap for your containers aren't metal, either, and that they don't have any metal parts inside.

You mix certain types of etchant with warm water. While stirring, chemical reaction heats up the etchant even more, so be sure to use a plastic tray that can hold *very* hot water. Processing trays for photography generally make ideal etching troughs. For safety, check the tray before you use it: Fill it with hot (150 to 180 degree) water. The tray shouldn't become soft or melt.

For best results, a tray should have ripples or ridges on the bottom. This texture allows the etchant solution to flow freely under the board while the board stews in the tray. Finally, you need two plastic or bamboo tongs to handle the board. Don't use your fingers! You can buy these tongs at any photographic or arts supply store.

You find etchant, whether ferric chloride or ammonium persulfate, in three popular forms:

✔ Liquid, not concentrated

✔ Liquid, concentrated

✔ Powder (sometimes this comes as a semi-glutinous paste)

You can get liquid, unconcentrated etchant at Radio Shack and most electronics stores. It comes in a plastic bottle ready for use. Just open the bottle, pour the etchant solution into a plastic (remember, never metal) tray, and you're ready to go.

You can use unconcentrated liquid etchant to make more than one board, depending on the size of the boards. The etching action reduces as you increase the surface area of the board.

For example, if the board measures 4 x 6 inches, with one side to etch, the board has 24 square inches of copper clad. Check the bottle for your etchant's recommended usage. Your particular solution may be able to etch up to 50 square inches of copper clad. This estimate assumes that you use the entire contents of the bottle. If you use less etchant, you also reduce the expected amount of coverage.

The size and number of boards that you make determines how long the etchant lasts before it just can't etch anymore. You need to throw out weaker etchants after you use them to make just one 2 x 3-inch board; you can use stronger etchants to make several large boards. Using weak etchant, you may have to wait ages for the etching to finish, and this weak etchant can lead to voids in the copper pattern.

Here are some tips to keep in mind when mixing and using etchants for making printed circuit boards:

- ✔ You must dilute concentrated liquid etchant before you use it. For best results, dilute the etchant with hot water; this addition increases the etching action. Typical dilution ratios are 2:1, 3:1, and 4:1. The higher the ratio, the longer the concentrate lasts. For best results, though, balance the thrifty use of the concentrate with your tolerance for longer etching times. The weaker the etchant, the longer it takes to remove the excess copper.

- ✔ You have to mix powder (or paste) etchant before you use it. One packet of powder etchant generally makes one or two quarts of unconcentrated etchant. You can mix the powder to make a smaller amount of liquid and then dilute the mixture when you're ready to use it.

Now that you're itching to etch . . .

After you go through all the preliminaries in the preceding sections, you get to actually etch your printed circuit board.

Follow these steps to etch the board:

1. **Pour the etchant into the plastic tray carefully, avoiding spills and splashes.**

   Pour enough etchant to create a pool at least ⅛-inch thick, preferably ¼-inch thick.

2. **Dunk the board into the tray and continually rock it back and forth.**

3. **Keep the board in the soup for 10 to 30 minutes (depending on the type and strength of the etchant) or until the etchant has removed all the excess copper. Keep that tray a-rockin — but gently!**

4. **Use the plastic or wooden tongs to lift the board out of the tray from time to time to check progress.**

The etchant removes the copper, starting from the edges and areas close to the resist. Large, open areas of copper can be stubborn and take 2 to 3 times as long to etch completely. You may want to agitate those areas of the copper that don't respond as quickly to the etchant. However, be sure that you don't over-agitate because you can undercut the copper under the resist. *Under-cutting* happens when etchant oozes under the resist and attacks the copper that you don't want to remove.

Etchant, be gone!

Diluted etchant solution that you use for hobbyist applications usually doesn't pose a serious threat to plumbing, but the etchant is a pollutant and a toxin. Take the exhausted etchant to a licensed recycler or approved chemical waste disposal site. If you live near an electronics manufacturer, you may be able to get them to dispose of it for you. Because they can reclaim copper from exhausted etchant solution, the company may not charge you for this service.

Final Prep and Drilling

If you've gone through the steps earlier in this chapter, you've almost created your first finished board. But before you turn off your work light for the night, you have one more phase to handle: you have to do final prep and drilling of the circuit board.

The etchant has completed etching when you can't see any traces of exposed copper. Assuming that you did the etching process correctly, the copper under the resist should remain intact. Still, the black resist for the traces and component pads remains, so first you have some clean-up duties to perform. After etching, rinse the board under cold running water for 15 to 20 seconds. Be sure to rinse the etchant from the back side of the board, as well.

After you clean the board, follow these steps to prepare and drill it:

1. **Remove the resist with lacquer thinner or thoroughly scrub the board with a non-metallic scouring pad and cleaner.**

   We regularly use Ajax® cleanser and green Scotchbrite™ scouring pads. When you've completely removed the resist, the copper should be bright and shiny, with no evidence of undercutting.

2. **If you find that the board was over-etched and it's missing some of the traces, you can repair the board by soldering short lengths of wire to bridge the gaps or by applying copper tape to the missing portions.**

   You can find a variety of copper tape and pads to make or repair PCBs. Check out the Appendix for a list of several online sites that carry printed circuit board–making products. The copper pieces have adhesive backing that you use to fix them to the board.

3. **Drill the board using a small 0.040" or 0.070" drill; we recommend the smaller drill for IC sockets and small resistors and capacitors.**

We also recommend a small drill press, like the one you see in Figure 12-7, or a drilling stand. For best results, keep most of the bit tucked inside the drill motor chuck. With only ¼ to ½ inch protruding from the chuck, you have less chance of breaking or bending the bit. If you're in doubt about how to use your drill or drill press, check the instructions that came with it!

4. **Position the drill exactly within the hole on each pad. If a component pad doesn't have a hole, drill in the approximate center of the pad.**

   Most components don't require precision drilling, but with some — most notably, ICs — you must drill the hole within about ⅟₅₀ of an inch of the proper spot. If the drill bit dances around the copper foil before it digs in, use a center punch to make a small dimple in the board. The dimple helps you aim the bit at the exact spot that you want for the hole.

5. **After drilling, inspect the copper for burrs and remove them with an emery board or fine steel wool.**

6. **Examine the back of the board for chips and cracks; remove broken pieces of epoxy and file away the rough spots.**

Drill from the foil (copper) side of the board to the back. You can prevent splitting and chipping the back of the board by placing a sheet of scrap wood underneath it, but don't use particle board because it quickly dulls the drill bit.

Figure 12-7:
Use a small drill press to drill the holes in your PCB.

One more cleaning duty: Thoroughly wash the board to remove pieces of copper, epoxy, wood, steel wool, and other contaminants. Use a non-metal scouring pad (such as 3M Scotchbrite™) and cleanser to thoroughly clean the copper foil.

You're now ready to solder the components to your custom-made printed circuit board. (You can get into the soldering groove by checking out Chapter 8.) If you don't plan to use the board for a while, place it in a plastic baggie and store it in a safe place where it can't get dirty.

PCBs R Us: Using a PCB Service

What if you don't want to get your hands dirty making a PCB with the methods that we describe in the earlier sections of this chapter, but you still want to permanently mount that circuit? Just have a company that makes PCBs for a living make one or two (or a hundred) for you.

Now you're a board designer

To get a PCB board made by a PCB manufacturer, you first need to generate the PCB layout. You can generate the layout with Computer Aided Design (or CAD) software (which we discuss in more detail in the section "Using CAD to Make Artwork," later in this chapter). Using CAD software, you generate data files (called Gerber files), which you then send to the PCB manufacturer. The manufacturer uses these files to give you a quote, and if you're okay with the price, they go ahead and make the board for you.

PCB design is full of rules, just like so many other things in life. Before you generate Gerber files, check out the manufacturer's design rules. Design rules insure that the PCB you want doesn't require features that the manufacturer's equipment and processes can't handle. The manufacturer checks your files to see that they meet the design rules (this once-over is called a *design rule check,* or DRC). Most manufacturers probably won't do the job until you correct any design rule errors.

Common design rules require that you maintain a certain minimum

✔ Trace width

✔ Space between traces

✔ Space between a trace and the outside of a board

✔ Pad size

The list above doesn't cover everything, so be sure to check the manufacturer's Web site for their complete design rules.

A word to the wise: Some PCB vendors offer the option of getting boards at a lower price by skipping the silk screen and solder mask steps. Silk screening puts ink letters and numbers on the board for specific purposes, such as marking the hole into which you should solder resistor R3. *Solder mask* is a green film that protects the traces on the board if you get a little sloppy and spill solder where you shouldn't. Generally, we recommend getting your board made with silk screening and the solder mask: These features are worth the few extra dollars that you spend because they help you keep track of components and protect your board.

PCBs: Everybody's doing it (But will they do it for you?)

There is no shortage of companies that manufacture PCBs; the tricky part comes when you have to find ones that take orders for anywhere from one to a handful of boards from a hobbyist for a reasonable price.

We point out some manufacturers that fit this criterion, but we suggest that you compare prices among manufacturers when you're ready to buy. You can do a search using Google or your favorite search engine for "printed circuit board manufacturer" and compare prices by visiting the Web sites that the search engine finds for you.

As of this writing, we recommend that you take a look at these three manufacturers:

- ✔ **Olimex:** To obtain a board at a low cost, try Olimex (www.olimex.com). The only drawback to Olimex is that they have longer shipping times than other manufacturers because they're not-so-conveniently located in Bulgaria.

- ✔ **AP Circuits:** If you can't wait the few weeks it takes to ship products from Europe, try AP Circuits (www.apcircuits.com). Although their prices aren't quite as low as Olimex, they're reasonable. And if you're in the US or Canada, you're looking at a much shorter shipping time.

- ✔ **Advanced Circuits:** If you're a college student, you may want to check out Advanced Circuits (www.4pcb.com). They waive their minimum order requirement for college students, allowing you to order one board for the same price per board that the rest of us pay when ordering three boards. Who says education doesn't pay off?

Using CAD to Make Artwork

You use Computer Aided Design programs to make layout art for PCB boards. These programs contain libraries of symbols for common components and templates, and they include tools for creating drawings. They provide several features that help turn your ideas into polished drawings.

You can buy sophisticated CAD programs for hundreds or thousands of dollars (and you may have to be a rocket scientist to use some of them); or, you can download freeware or shareware CAD programs. Some of these are pretty simple to use. Do a Google search for "CAD software for PCB design," and you can find several software packages.

The program that we recommend is Eagle Light from CadSoft. You can download this program for free at www.cadsoft.de.

What you can do with Eagle Light CAD

Of course, you pay a price for getting things free: The free Eagle Light CAD program has a few limitations. You can't use it to draw a board larger than 4 inches by 3.2 inches or one containing more than two layers. If you need to make a board that exceeds these limitations or are making boards for a profit, go ahead and fork over the money for the full version of the software (about $200 for the standard version and $400 for the professional version at the time of this writing).

You can use Eagle Light software to produce board layout drawings and data files. These Gerber files contain the data that PCB manufacturers need to make your custom board. You can also print out the layout drawing and use it to make the PCB yourself with one of the methods that we cover in the earlier sections in this chapter.

Getting to work designing a board

We suggest that you read the tutorial for Eagle Light on the CadSoft Web site (www.cadsoft.de) and then do the low-tech thing: Print out the page of the tutorial that gives you the tool button functions and tape it next to your computer monitor. The tutorial, along with the demonstration schematics and layout drawings they include with the download, give you enough to get started generating your own PCB layout drawings and files.

We won't kid you: You may have to stumble around a bit the first few times that you use the software before you get it right. But stick with it, and you can soon become an Eagle Light master.

To give you an idea of what you can do with a CAD program, we outline the major steps for using Eagle Light here:

1. **Enter your circuit into the CAD software.**

   Entering your circuit into the software involves steps such as placing symbols for all the components from your circuit into your onscreen layout and then drawing lines that represent the wires between the appropriate pins, placing junctions in the appropriate places, and attaching +V and ground symbols to the appropriate pins.

2. **Next, you run an automated electrical rule check (ERC) by clicking the ERC tool button.**

   This check catches problems that you may have missed, such as failing to attach ground to the correct pin of an IC.

3. **Correct any errors and run the ERC again.**

Figure 12-8 shows a schematic drawn in Eagle Light. (Credit where it's due: The drawings in this section come from a barometer project constructed by Philip Gladstone.)

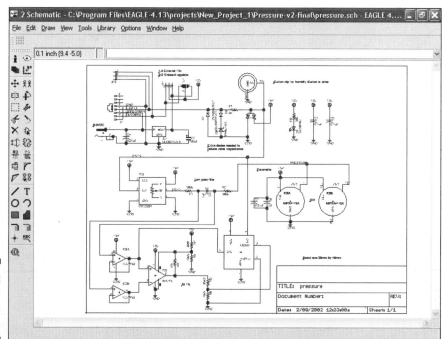

Figure 12-8:
A schematic
drawn in
Eagle Light.

Figure 12-8 shows the complete schematic, but that's a whole lot of schematic to take in at once. To show you the details, we zoom into a portion of the schematic in Figure 12-9.

After you correct any errors, start the process of generating the layout drawing (the layout drawing is the actual artwork that will become your printed circuit board):

1. **Click the Switch To Board tool button.**

   Eagle Light opens a board window that includes the symbols for each of the components that you specify in the schematic.

2. **Click and drag to place the symbols in the correct position on the layout drawing.**

3. **Click the Autorouter button; the program draws traces that correspond to the connections that you indicate in the schematic.**

If the Autorouter can't place all the traces, move some of the components a bit and try the Autorouter again. For example, the Autorouter may not position components placed too close together just right. If this second attempt doesn't work, you may have to route some of the traces manually.

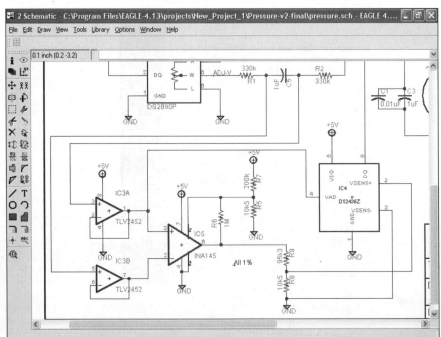

Figure 12-9:
A portion
of the
schematic.

Figure 12-10 shows a layout drawing with the traces, pads, vias (vias connect traces on different layers), and component numbers for the top layer of the board, based on the schematic in Figure 12-8.

You can see the layout drawing showing the traces, pads, and vias for the bottom layer of this same board in Figure 12-11. Note that because this is the bottom of the board, the text shown in Figure 12-10 is reversed in Figure 12-11.

After you finish routing the board, run a design rule check (DRC) to make sure you've properly drawn the board and avoided violating any design rules. Eagle Light then asks you to check the design rules against the manufacturer's.

The default design rules built into Eagle Light often work just fine, but go ahead and check against the PCB manufacturer's rules and make any necessary changes. For example, if the PCB manufacturer has a minimum trace width of 10 mils and the default in Eagle Light is 8 mils, change the minimum width to 10 mils.

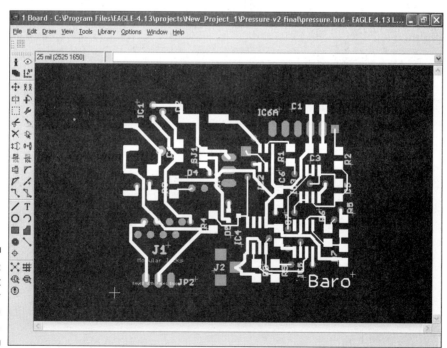

Figure 12-10: Layout drawing for the top layer.

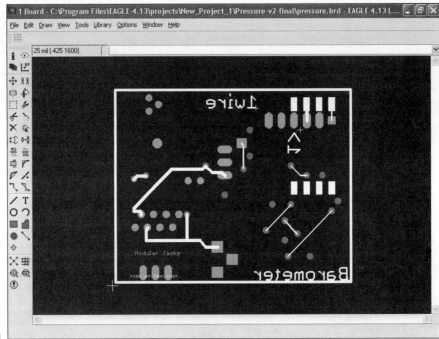

Figure 12-11:
Layout
drawing for
the bottom
layer.

That's it. Now you're ready to generate Gerber files from the layout drawing and send them to the manufacturer. Use the printed circuit board software to generate the proper files for you. You need to discuss with the PCB manufacturer how they prefer to receive files. Some have an automated process where you submit the files via their Web page, and others are perfectly happy receiving the files through regular e-mail.

Using the Autorouter feature, rather than manually routing the traces, minimizes the chance of creating any design rule violations.

Figure 12-12 shows the top side of a board made from the layout drawing that we show you in Figure 12-10.

You can see the bottom side of the finished board, whose layout drawing you can see in Figure 12-11, in Figure 12-13.

Finally, we show you the board with all the components in place in Figure 12-14.

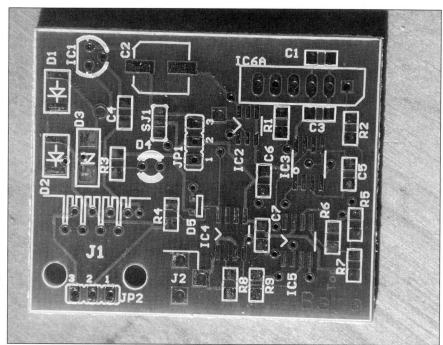

Figure 12-12:
The top side
of the PCB.

Figure 12-13:
The bottom
side of the
PCB.

Figure 12-14:
The completed board.

Chapter 13

The Exciting World of Microcontrollers

In This Chapter

▶ Exploring how microcontrollers work

▶ Getting into a microcontroller's guts

▶ Running down microcontrollers for students and hobbyists

▶ Taking a closer look at some microcontrollers

▶ Going on a microcontroller-info hunt

*M*icrocontrollers are the Eighth Wonder of the World. What makes microcontrollers so special? Simply put, they're programmable circuits. Just like your home computer, you can program them to do whatever you may want them to do.

Though they may look like ordinary integrated circuits, they have much more to offer; in fact, microcontrollers have become *the* way to make the best and brightest electronic products. Here's just one example: if you have a car built within the last 10 years, odds are it uses not one, but a half dozen or more microcontrollers. Each microcontroller is dedicated to taking care of some important facet of your driving experience, from the brakes to the electronic ignition to the air bag system.

We focus on the microcontrollers for the hobbyist in this chapter. Here, you can discover what microcontrollers are and what they do. In the section "Getting to Know the BASIC Stamp 2," near the end of this chapter, we give you two nifty hands-on demonstrations that you can try yourself to get a peek at the power of microcontrollers.

So, How Does It Work?

A microcontroller is an integrated circuit chip, which is usually mounted on a mini-PCB that includes other components in circuits that interface the micro-controller to your computer, motors, or switches. When you're programming a microcontroller, you place it on a development board that allows the microcontroller to interface with your personal computer. Once it's programmed, you mount the microcontroller into a socket on your electronic device.

Unlike traditional circuits, you don't need to swap wires around or pull out resistors or capacitors and replace them with some other part to change a microcontroller's function. Instead, you just alter a couple of lines of programming code. The microcontroller seems to take on a different personality, instantly changing what your project actually does. You can program a single microcontroller to do any of thousands, of different jobs!

Most microcontrollers are designed for use in commercial products, and you may find it a bit difficult to program these little guys. Fortunately, some versions of microcontrollers are specifically for the hobbyist, so you get everything that you need to run the microcontroller on one small circuit board. You can easily program these hobby-friendly microcontrollers, and they don't drain your wallet.

What's Inside a Microcontroller?

Originally, microcontrollers were designed to provide a way for a personal computer to communicate with electronic gadgets in the outside world. We still use them for that purpose, and more.

Here are the parts of a typical microcontroller:

✓ **Small computer:** This computer sits at the heart of the microcontroller. This built-in computer isn't as powerful as the one on your desk, but microcontrollers don't need a ton of horsepower. You expect your desktop computer to do several big jobs at once, such as browsing the Internet, calculating spreadsheets, and fending off viruses. The typical microcontroller does a single job.

✓ **Non-volatile memory:** The microcontroller stores the program that runs on its computer in non-volatile memory. This memory sticks around when you turn off the power. The non-volatile memory comes to life the moment that you connect the batteries and flip the switch.

✓ **Input/output ports:** These connections on a microcontroller allow it to communicate with the real world, running things like lights, motors, relays, sensors, switches, liquid crystal displays, and even other microcontrollers. These input/output ports, also called I/O ports, provide information that allows the chip to control your project. A microcontroller program may light up an LED when you flip a switch or run a motor when a sensor detects someone walking by, for example.

You can see a good example of a microcontroller in action in the brains of the LEGO Mindstorms robot construction set. The yellow brick that you can see in Figure 13-1 contains a small microcontroller chip that can display messages on a liquid crystal display (LCD), react to switches and other sensors, and run up to three motors at the same time.

As with all microcontrollers, you program the LEGO Mindstorms' brain by sending it programming instructions. First, you create these instructions on your personal computer, and then you transmit them to the Mindstorms brain by using an infrared link (most microcontrollers that hobbyists work with use a wired link that you connect to your PC's serial or USB port). After you send instructions to the microcontroller, the instructions reside in non-volatile memory until you replace them with new instructions. LEGO Mindstorms gives you a good example of a microcontroller's ability to play multiple roles; you just have to modify the software. By changing a few lines of a program, a LEGO Mindstorms robot can do the following:

- ✔ Search for the brightest light in the room, such as a flashlight, and move toward it.

- ✔ Find the brightest light and, instead of approaching it, move away from it.

- ✔ React to bumper switches mounted on its sides so that when the robot hits an obstacle, it backs up and goes the other way.

- ✔ Sense a black line on a piece of white construction paper and follow it.

The LEGO Mindstorms, like the little robot in Figure 13-2, can also follow a combination of several of these functions. This little 'bot can follow a bright light *and* back up from obstacles that it bumps into.

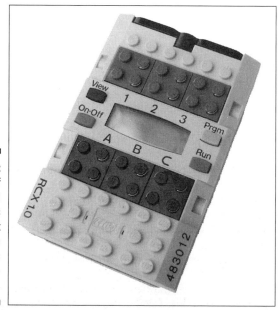

Figure 13-1:
The brain of the LEGO Mindstorms robot construction kit contains a micro-controller.

Figure 13-2:
Software
residing in
this LEGO
Mindstorms
controls
what the
robot does.

You have a nearly endless list of possibilities when you start reprogramming a microcontroller. Writing a new program and downloading it to the microcontroller takes much less time than rebuilding a circuit, which is precisely why electronics folks find microcontrollers so darn useful.

Discovering Microcontrollers for Hobbyists

You can choose from hundreds of microcontrollers, but only a handful give first-time experimenters what they need. Some manufacturers simply don't sell certain brands of microcontrollers to the general public; you can get other brands from a variety of online and retail stores.

In general, a hobbyist can buy microcontrollers that fall into one of two main categories: those with an embedded language interpreter and those without.

An *embedded language interpreter* is a program that runs inside the microcontroller. It allows you to write your programs with an easy-to-use language. You download your program to the microcontroller, and the interpreter converts it to the type of instructions (called *assembly language code*) that the microcontroller understands.

Microcontrollers that have an embedded interpreter are easier to figure out and use effectively. If you're just starting out, opt for one of these user-friendly microcontrollers. Good examples include the BASIC Stamp and OOPic microcontrollers, both of which we discuss in the section "Microcontrollers That Stand Out from the Rest" later in this chapter.

The most common language used for embedded interpreters is BASIC. Good news for computer geeks: if you've ever played around with writing computer programs in BASIC, you're already well on your way to programming a microcontroller! Of course, if you've never programmed in BASIC, you get to explore a whole new language. But never fear. BASIC isn't rocket science. You can master it soon enough.

If you're serious about figuring out how to use microcontrollers, check out *Beginning Programming For Dummies,* by Wallace Wang (Wiley Publishing, Inc.).

Some microcontrollers, namely the BASIC Stamp (which we cover in more detail in the section "Introducing the BASIC Stamp" later in this chapter), come with fairly thorough printed documentation. Often, you can find everything that you need to complete the project at hand in the user manual that comes with the microcontroller kit.

You create the program for the microcontroller in a program editor. You also have to get a special programming hardware module that links your PC and the microcontroller chip.

You program microcontrollers that don't contain an embedded interpreter by using either assembly language or a high-level language.

- ✔ **Assembly language:** The hardest to understand and use, we don't recommend assembly language if you're just starting out. It's hard to read assembly language programs and harder to fix them when they don't work.

- ✔ **High-level languages:** 98 percent of all computer programs are created with these languages. They support a rich variety of options and often make up part of a more elaborate development platform that makes it a lot easier for you to *debug* (find and fix problems). For programming microcontrollers, the three most common high-level languages are BASIC, C, and Java. Most beginners use BASIC because you can master it most easily.

How much is that microcontroller in the window?

Going just by retail prices, microcontrollers vary from about 50¢ to over $100. Why the huge disparity? There are several reasons:

✔ **Microcontrollers with an embedded interpreter cost the most.** The cost of the microcontroller includes the built-in interpreter, as well as an easy-to-use connection to your PC. (For most microcontrollers without an embedded interpreter, you have to buy a separate hardware-programming module that provides the electrical link between the controller and your PC.)

✔ **Features, such as the amount of memory or the number of I/O ports, affect cost.** The least expensive microcontrollers have just three or four I/O ports. The more elaborate microcontrollers have 30 or 40 ports. The more I/O ports, the more things your microcontroller can control — and the more money you have to fork over to buy it.

✔ **The ability to re-program the microcontroller over and over again increases the cost of the chip.** You can program the cheapest controllers once; therefore, people call these controllers *OTP*, which stands for one-time programmable. For a few dollars more, you get a microcontroller with erasable memory: You program it, and then you can erase the existing program and record another in its place. Most microcontrollers use a type of memory called Flash, the same kind that your digital camera or MP3 player uses. You can erase Flash memory and fill it with a new program a thousand or more times.

PC calling microcontroller: Come in, please!

If you purchase a microcontroller without an embedded interpreter, you need to also buy (or make) a hardware-programming module. The hardware module provides a physical link between your computer and the microcontroller.

You can use most commercial modules to program several microcontrollers that come from the same manufacturer, so look for that feature. It doesn't make sense to get a hardware module that programs just one specific microcontroller because you lock yourself into just that chip.

The price tag for hardware modules runs from just a few dollars to $100 or more. You can also build your own module, but most first-timers opt for a ready-made one to save time and hassle. Take a whack at building your own module after you've gained a little experience with microcontrollers.

Figure 13-3 shows a typical commercial hardware-programming module. This particular module also provides built-in buttons and lights to help in developing applications You don't have to use the added development features just to program a microcontroller, but they can be nice to have, and don't usually add much to the price of the module. You plug the chip into a socket in the module and attach a cable to your PC. Most, but certainly not all, modules come with at least one editor that you can use to write your programs. If the module you buy doesn't have a programming editor, you need to get one.

Figure 13-3:
You can use a hardware module like this one to program micro-controllers.

TIP

Explaining how to select a programming module and programming editor for a given microcontroller goes way beyond the scope of this book. You can choose from dozens of options, and the marketplace changes all the time. Your best bet is to contact the manufacturer of the microcontroller that you want to use and ask for recommendations.

Microcontrollers That Stand Out from the Rest

Of the dozens of brands of microcontrollers, two stand out as ideally suited for hobbyists. The following sections give you a rundown of these microcontrollers. Only you can decide which one fits you and your projects.

Introducing the BASIC Stamp

The BASIC Stamp is one of the best-known and most widely used microcontrollers for students and hobbyists. That popularity doesn't come from its lightning speed or a gaggle of features; it gets its popularity because it was one of the first microcontrollers to include an embedded interpreter.

The BASIC Stamp derives its name from its built-in BASIC language and its postage-stamp size (well, a kind of large stamp). You can find out more about the BASIC Stamp by visiting www.parallax.com.

The BASIC Stamp comes with some of the best documentation that you can find for a microcontroller. If you're just starting out, you can't go wrong with the BASIC Stamp because of all the tutorials, how-to's, references, and project ideas that you can easily find for it.

Tasting different flavors

You probably think a BASIC Stamp is . . . well, basic. But the BASIC Stamp comes in varieties — 1, 2, SX, and a slew of others. These versions differ in features and, in some cases, the embedded programming language. Parallax sells a version of the BASIC Stamp, called the Javelin, which has the Java programming language embedded rather than BASIC, for example.

The BASIC Stamp 2, or BS2, is one of the most popular of the gang and the one that we recommend when you're just starting out. The BS2 comes on a single 24-pin chip, as you can see in Figure 13-4. This one chip is actually a carrier: it contains a lot of little integrated circuits and other components, such as extra memory and a voltage regulator.

Figure 13-4:
The venerable BASIC Stamp 2 all-in-one micro-controller.

Adding a development board

Although you can use the BS2 chip as-is, most beginners enhance it with a project board. Parallax sells several boards that you can choose from. One board, cleverly named the Board of Education (or BOE), has a larger capacity voltage regulator built in, as well as extra connectors for hooking things up to it. The BOE, which you can see in Figure 13-5, also includes a little solderless breadboard right on it so that you can try out different circuits. Very handy!

Figure 13-5:
The Parallax Board of Education.

You can buy the BOE with or without a BS2 chip. You can also find versions that you connect to your PC using a serial or USB cable. If your computer lacks a serial port — if you have a late-model laptop, for example — you have to get the USB version.

Don't forget the programming software

You have to get specialized software for programming any version of the BASIC Stamp. You can get the software for free. It comes included as part of a BASIC Stamp starter kit. You can also download it from the Parallax Web site. You can find several versions of the software for use with MS-DOS, Windows 98 or later, Macintosh, or Linux.

You can run the MS-DOS editor under Windows 95 or 98. This great option lets you dedicate an old, otherwise useless PC just to programming the BASIC Stamp.

Kicking it up a notch with PICMicro and Atmel AVR

No discussion of microcontrollers would be complete without mentioning the PICMicro series, from Microchip, and the AVR series, from Atmel. The PICMicro and AVR families lack an embedded language interpreter. Instead, you program them using a special hardware module (see the section "PC calling microprocessor: Come in, please" in this chapter) and a custom programming editor.

This programming approach has one advantage: you can choose from among a number of different programming languages and development platforms. This is useful if you're already familiar or comfortable with a particular programming language. You don't have to learn yet another language if you pick up an interest in microcontrollers. But this route is a lot more complicated. We don't recommend it if you're just starting out with microcontrollers or have never done much programming. And, depending on the microcontroller chip, you may need to use extra circuitry to make it work. This extra circuitry includes a crystal oscillator, some

capacitors to make the oscillator function, and a voltage regulator.

If you want to try your hand at the PICMicro or AVR microcontrollers, check around for a starter kit that includes everything that you need: a sample microcontroller chip, a programming module, software, and cables to connect to your PC. Many online retailers, such as Digikey or Jameco, sell these starter kits for PICMicro or AVR microcontrollers. The starter kit makes figuring out how to use these beasties a lot easier.

You can read more about the PIC Micro and AVR controllers at the Web addresses provided here. While neither company sells to consumers directly, the sites provide lists of online distributors you can contact. The Web sites contain product information, datasheets, and detailed training materials on using the products.

✔ **PICMicro:** www.microchip.com

✔ **AVR:** www.atmel.com

If you happen to already have a BASIC Stamp and programming editor, check the version of the software. If it's kind of dated, be sure to update the software with the latest and greatest available at the Parallax site. Newer versions of the BASIC Stamp software include some handy additional features that you will definitely want!

Introducing the OOPic

The OOPic — pronounced ew-pik — is a relative newcomer to the exciting world of hobby microcontrollers. Even so, it's catching on fast. The OOPic uses a radically different method of programming, involving objects.

The idea behind this type of programming is that you use these objects, instead of writing lots and lots of code to program common tasks. Because most experimenters use microcontrollers for the same purposes, such as operating motors or reading switch settings, the OOPic makes your job easier by letting objects do much of the work for you. Using objects, you avoid the traditional method of programming microcontrollers where you have to write several lines of code for each of these tasks. The result: you really save a lot of time.

With the OOPic, they built the functionality for working with real-world devices into the chip. You just tell the OOPic what you connected it to and then give it simple commands. The OOPic figures out how to control the specified device all on its own.

This approach to programming isn't new or novel to computer programmers, but it is unique to microcontrollers. It may take you a little while to get used to the programming style of the OOPic, but after you get the hang of it, you find that it simplifies a lot of mundane tasks. Read more about the OOPic on the Web site at www.oopic.com.

Going all-in-one or single chip

Like the BASIC Stamp, you can find the OOPic in a couple of different versions. You can buy the OOPic already connected to a carrier board (see Figure 13-6) or as a 24-pin chip, just like the BASIC Stamp 2. Our favorite version of the OOPic is the OOPic R, the one that you can see in Figure 13-6. The carrier board comes equipped with a speaker, several switches, some LEDs that act as indicator lights, and a whole mess of connector pins that you use to attach components.

Figure 13-6:
The OOPic R provides an all-in-one develop-ment platform.

Speaking the right language

You can set the OOPic's programming software to use any of three popular languages: BASIC (recommended for beginners), C, or Java. You can get the programming software as a free download from www.oopic.com. The OOPic Webmaster updates the site from time to time with new features.

Getting to Know the BASIC Stamp 2

Now it's time to take a closer look at how you can use one microcontroller — specifically, the BASIC Stamp 2 — to create some simple electronics projects. Mind you, the power of microcontrollers goes way beyond the basic examples that we provide in the following sections. But the beginner's projects should give you an idea of the power that lurks under the hood of one of these babies. When you're done with these, check out Chapter 15, where you explore how to use the BASIC Stamp 2 to build a small, intelligent robot.

Step 1: Making the circuit

Although the BASIC Stamp is self-contained and doesn't need anything else to perform its function, you still need to connect things to it. What you connect to the BASIC Stamp depends on what kind of circuit you want to build.

Suppose, for example, that you want the program in the BASIC Stamp to flash an LED. This circuit always works well as a demonstration because you only need two components — a resistor and an LED — and you can easily tell if the connection works or not.

Take a look at the diagram in Figure 13-7. To build this circuit, you connect a resistor and LED to the BASIC Stamp as Figure 13-7 shows you. You can plug things into the BASIC Stamp a whole lot more easily if you have an experimenter's board, such as the Board of Education (BOE). The BOE also provides a convenient way to connect the BASIC Stamp to your PC by using a serial or USB cable, and to plug in power. If you use the BOE, you'll find that I/O Pin 0 is labeled as P0 in the column of labels next to the solderless breadboard.

Figure 13-8 shows you the blinking LED circuit on a BOE.

Step 2: Programming the darned thing

With the circuit all hooked up, you're ready to program the BASIC Stamp to flash the LED. We assume that you already have the BASIC Stamp programming software installed on your computer, all set up and ready to go. (If you don't, you have to do that part now. We can wait . . .)

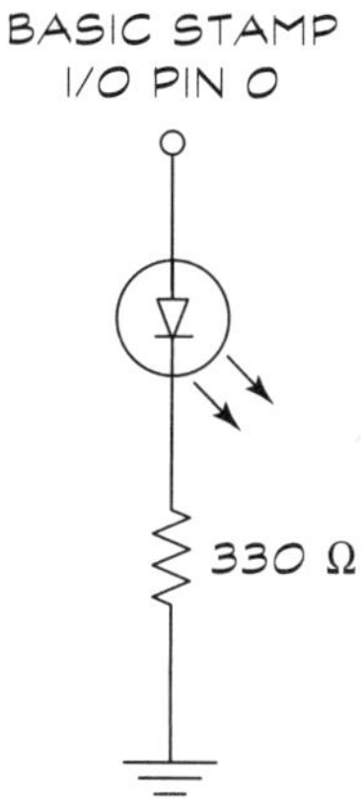

BASIC STAMP
I/O PIN 0

330 Ω

Figure 13-7:
Follow this
wiring
diagram to
create the
flashing LED
demonstra-
tor.

Figure 13-8:
The BOE
provides a
convenient
way to
make up
quick test
circuits.

Ready? Start the BASIC Stamp programming editor and type in the following short program. When you're done, the editor window should look like the one in Figure 13-9.

```
' {$STAMP BS2}
loop:
  HIGH 0       ' pin 0 high (the LED turns on)
  PAUSE 250    ' wait 250 milliseconds
  LOW 0        ' pin 0 low (the LED turns off)
  PAUSE 250    ' wait 250 milliseconds
  GOTO loop    ' loop forever
```

Here's how the LED flasher program works, line by line:

✔ **Line 1:** Tells the editor what kind of BASIC Stamp you're using. This example uses BASIC Stamp 2, so the line reads (`$STAMP BS2`).

✔ **Line 2:** This is what you call a label. You use it here and later in the program, on the last line, to create a never-ending loop.

✔ **Line 3:** `HIGH 0` turns I/O pin 0 on (makes it high). Because the LED connects to I/O pin 0, this line turns the LED on.

✔ **Line 4:** `PAUSE 250` makes the BASIC Stamp pause for 250 milliseconds. Remember that a millisecond is one thousandth of a second, so 250 milliseconds is 250/1000, or a quarter of a second.

✔ **Line 5:** `LOW 0` turns I/O pin 0 off (makes it low). This turns the LED off.

✔ **Line 6:** `PAUSE 250` makes the BASIC Stamp pause again for another 250 milliseconds.

✔ **Line 7:** `GOTO loop` tells the BASIC Stamp to go to the label named loop. This command causes the program to repeat over and over again, until either you turn the BASIC Stamp off or you reprogram it to do something different.

Commanding all BASIC Stamps

Take a look at the LED demonstrator program in the section "Step 2: Programming the darn thing." The program uses words such as `PAUSE`, `HIGH`, and `LOW`. These words are commands, also called *programming statements,* which tell the BASIC Stamp what to do. The BASIC Stamp provides several dozen of these programming statements. The way that you use these statements in your code determines what your program does. To use the BASIC Stamp, or any microcontroller, you have to master these programming statements.

One of the best ways to get comfortable using programming statements is to try out demonstrator programs, such as the one in the section "Getting to Know the BASIC Stamp 2." Look for short examples in the BASIC Stamp documentation and check out books, magazines, and Web sites that provide information on the BASIC Stamp. The more hands-on experience you have, the faster you discover how things work.

After you build an example circuit for the BASIC Stamp, don't be afraid to experiment by changing some of the programming or connecting different components. You can build very elaborate programs this way, one little piece at a time.

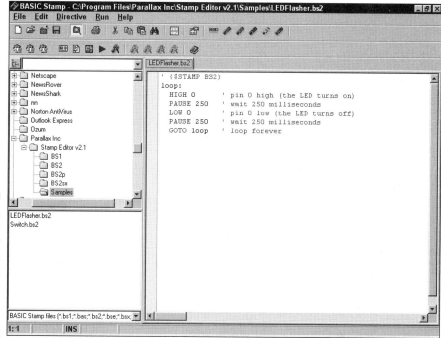

BASIC Stamp - C:\Program Files\Parallax Inc\Stamp Editor v2.1\Samples\LEDFlasher.bs2

```
' {$STAMP BS2}
loop:
    HIGH 0        ' pin 0 high (the LED turns on)
    PAUSE 250     ' wait 250 milliseconds
    LOW 0         ' pin 0 low (the LED turns off)
    PAUSE 250     ' wait 250 milliseconds
    GOTO loop     ' loop forever
```

Figure 13-9:
The LED
flasher
program in
the BASIC
Stamp
program
editor.

The BASIC Stamp software editor treats any text after an apostrophe as a comment. You make comments just for yourself; the BASIC Stamp ignores them and doesn't process them as instructions. Line 1 of the LED program falls into the comment category. You may want to get into the habit of adding at least a few comments to your programs to remind you of why you wrote what you wrote. Then, down the road, if you revisit a program, the comments give you a handy reminder of your intentions.

Step 3: Let 'er rip!

Okay, budding programmers: you're now ready to upload and try out the program:

1. **Connect a serial or USB cable to the Board of Education and your PC.**

   The type of cable depends on the version of the BOE that you have.

2. **Apply power to the BOE by plugging the wall transformer into the BOE power jack.**

   Alternatively, you can attach a 9-volt battery to the battery terminals in the upper-left corner of the BOE.

3. **In the BASIC Stamp editor, press** `Ctrl+R` **(the Ctrl key and the R key at the same time).**

This keystroke combo runs your program and downloads it to the BASIC Stamp.

If you have everything set up right, the LED should begin to flash twice each second. If the BASIC Stamp editor displays an error on your computer screen, locate and fix the problem (maybe a typo in your code?) and try again.

Making changes made easy

Here's where the power of microcontrollers really shines! Make the following changes in the program that we describe in the section "Step 2: Programming the darned thing" earlier in this chapter:

- Change Line 4 to `PAUSE 100`
- Also change Line 6 to `PAUSE 100`

Now, run the program (press `Ctrl+R`). What happens when you run the program this time? Your computer downloads the changes into the BASIC Stamp, and the LED now flashes much more quickly. Instead of pausing 250 milliseconds each time that the LED turns on and off, the BASIC Stamp pauses for only 100 milliseconds.

Change the program again with these adjustments:

- Change Line 4 to `PAUSE 1000`
- Also change Line 6 to `PAUSE 1000`

You can probably guess what happens. The LED flashes relatively slowly . . . once every other second. The BASIC Stamp now pauses 1000 milliseconds, or one full second, each time the LED turns on or off.

As you can see, by simply changing a line or two of programming code, you alter the behavior of your circuit.

Adding a switch to the mix

In this section, you can experience the versatility of the BASIC Stamp. You use the LED for the next demonstration, so make sure that you leave it hooked up, as we describe in the section "Getting to Know the BASIC Stamp 2" earlier in this chapter. Connect a switch to the Board of Education, following the schematic in Figure 13-10. You can use any ordinary switch, but a momentary pushbutton does the job the best. Be sure to add the 10K ohm resistor as Figure 13-10 shows you. If you use the BOE, you see that I/O pin 1 is labeled as P1 in the column of labels next to the solderless breadboard.

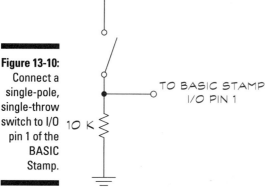

Figure 13-10:
Connect a single-pole, single-throw switch to I/O pin 1 of the BASIC Stamp.

What purpose does the 10K ohm resistor serve? It functions as a pulldown, which means that when the switch is not closed, the signal input feeding into the BASIC Stamp is 0 volts, or low. The resistor keeps the input to the BASIC Stamp from varying (called floating), which can cause the BASIC Stamp to give you erratic results.

Now, enter the test program:

```
'{$STAMP BS2}
OUTPUT 0        ' set pin 0 as output (for LED)
btn    VAR    Byte           ' define "btn" as a variable
loop:
   BUTTON 1,0,255,250,btn,0,noSwitch                ' check
            switch
   OUT0 = btn           ' turn LED on if switch was triggered
   PAUSE 150            ' wait 150 milliseconds
   OUT0 = 0      ' turn LED off
noSwitch: GOTO loop   ' repeat loop
```

Here's how this program works:

✔ **Line 1:** Tells the editor what kind of BASIC Stamp you're using — in this case, the BASIC Stamp 2. (The computer doesn't take any action regarding this comment line because of the apostrophe that begins it.)

✔ **Line 2:** OUTPUT 0 tells the BASIC Stamp to use I/O pin 0 as an output. You should have the LED connected to I/O pin 0 (if you don't, be sure to go through the steps in the section "Getting to Know the BASIC Stamp 2" earlier in this chapter).

✔ **Line 3:** This line, which reads `btn VAR Byte`, tells the BASIC Stamp to set aside a little bit of memory for a variable named `btn`. A *variable* is a temporary holding area for data. After the variable has been created, the BASIC Stamp can stuff data into it and then later come back and check to see what's in it.

✔ **Line 4:** The line `loop:` sets up a repeating loop, exactly as in the LED example in the section "Step 2: Programming the darned thing" earlier in this chapter.

✔ **Line 5:** This line starts with the `BUTTON` programming statement and tells the BASIC Stamp to check the state of the switch connected to I/O pin 1. The `BUTTON` statement requires a bunch of additional options, which you can find in the BASIC Stamp documentation that came with your BASIC Stamp kit.

✔ **Lines 6 through 8:** These lines turn the LED on, tell the BASIC Stamp to wait 150 milliseconds, and then extinguish the LED again.

✔ **Line 9:** Tells the BASIC Stamp to repeat itself, starting from the `loop:` label. This repeat goes on forever or until you unplug the BASIC Stamp or upload a new program.

See the `noSwitch` label in lines 5 and 9? When used with the `BUTTON` statement (as in Line 5), this label creates what is known as a branch. Should the switch not be depressed (the `noSwitch` part of the code), the BASIC Stamp jumps from the `BUTTON` statement line and goes all the way down to the last line, missing all the instructions for lighting the LED. But if the switch is depressed, the BASIC Stamp performs all the steps.

Here's what should happen when you run the program:

✔ When the switch isn't depressed, the LED stays off.

✔ When the switch is depressed, the LED blinks on briefly, and then turns off again.

Uh oh! If your BASIC Stamp doesn't respond this way, double-check how you have everything wired together, and verify that you typed the program in exactly as you see it in this section.

Where to Go from Here

This chapter touches on just the basics of the BASIC Stamp. You can find a whole lot more to do with the BASIC Stamp, or any other microcontroller that you want to try, for that matter. Check your neighborhood bookstore for references on using microcontrollers. Try Google or another online search engine to browse for goodies about your microcontroller of choice.

Chapter 14

Great Projects You Can Build in 30 Minutes or Less

- -

In This Chapter

▶ Stocking up on project supplies

▶ Creating unique light blinkers and flashers

▶ Exploring the smashing personality of piezoelectrics

▶ Seeing in the dark with an infrared sensor

▶ Rigging a couple of alarms

▶ Finding your way with your very own portable electronic compass

▶ Creating your own amplifier

▶ Testing for water

- -

*G*etting up to speed on electronics really pays off when you get to the point where you can actually build a project or two. In this chapter, you get to play with several fun, entertaining, and educational electronics gadgets that you can build in half an hour or less. We selected the projects for their high cool factor and their simplicity. We've kept parts to a minimum, and the most expensive project costs under $15 or so to build.

We've given you some detailed procedures for the first project, so work through that first. Then, you should be able to follow the circuit schematics and build the rest of the projects on your own. Check back to Chapters 6 and 7 if you need a little help reading or understanding the schematics.

Getting What You Need Right Off the Bat

You can build all the projects in this chapter, except for the electronic compass, on a solderless breadboard. Of course, feel free to build any of the projects on a regular soldered circuit board, if you want to keep them around.

We cover all the parts you use in these projects, such as transistors, integrated circuits, capacitors, and even wire in Chapters 4 and 5. There's more detail about breadboarding and building circuits in Chapters 11 and 12. If you get stuck on any of these projects, hop to one of those chapters to help you through.

With one exception (that pesky but worthwhile electronic compass project, again), you can find the parts that you need to construct the projects in this chapter at any electronics store or online retailer. If you don't have a well-stocked electronics outlet near you, check out both Chapter 17 and the Appendix for some mail-order electronics parts suppliers.

Unless we tell you otherwise:

- All resistors are rated for ¼ or ⅛ watt and 5 percent or 10 percent tolerance.
- All capacitors are rated at a minimum of 25 volts. We note the type of capacitor that you need (disc, electrolytic, or tantalum) in the parts list for each project.

Creating Cool, Crazy, Blinky Lights

The first project that one of us ever built was a light that blinked on and off. That's all it did, but that was enough. The project involved soldering together all the transistors, resistors, and diodes. Start to finish, the whole project took two days and cost a pricey $17 in parts. Today, thanks to one specific integrated circuit, making your first blinky light project doesn't take you more than a few minutes and costs less than two bucks. The special ingredient that makes the blinky light circuit easy to build is the LM555 timer IC. This particular chip is to electronics what milk is to cookies. It's the cornerstone of many projects that you build, including several in this chapter. You can use the 555 in a variety of ways, but the most important use is to provide pulses at regular intervals, sort of like an electronic metronome. Throughout this chapter you see several ways to use this feature to produce a number of cool effects.

You can buy the LM555 at most any electronics store. We like to buy them in packs of 15 or 20, to save money. Don't worry about overstocking; you use them up soon enough. The price for one may be anywhere from 75¢ to $1.50, but when you purchase in quantity, the price goes down to less than half that per circuit. Online retail stores that sell the 555 by the tube tend to offer the best price. The tube is 18 to 24 inches long (depending on the source) and contains as many as a couple dozen chips.

Taking a closer look at the 555 flasher

You can see the schematic of the blinky light project in Figure 14-1. This figure shows you how to connect a 555 timer IC to an LED. By turning variable resistor (potentiometer) R1, you change the rate of blinks from a slow waltz to a fast samba.

If you need a quick refresher course on reading schematics, head back to Chapter 6.

This circuit provides a useful demonstration of how you can use the 555 as an *astable multivibrator*. That's just a fancy term for a timer that goes off (not turns off, but goes off like an alarm clock) over and over again, forever (or until it runs out of juice to power it).

This circuit also makes a handy piece of test equipment. Connect the output of the 555 (pin 3 on the chip) to some other project and use this circuit as a signal source. You see how this works in several of the other projects in this chapter that are built around the 555 chip.

It's easy to build the LED flasher circuit. Use the schematic you see in Figure 14-2 as your guide. Note that we added a bit more space between components so that it's easier for you to see where all the parts go. You should usually build in a little bit of space, rather than squeezing things together, so you can see what you're doing.

Follow these steps to build the circuit:

1. **Collect all the components you need for the project ahead of time. See the parts list below for a rundown of what you need.**

   There's nothing worse than starting a project, only to have to stop halfway through because you don't have everything at hand!

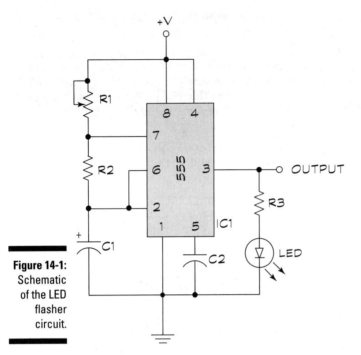

Figure 14-1:
Schematic
of the LED
flasher
circuit.

2. **Carefully insert the 555 timer chip into the middle of the board.**

   The IC should straddle the empty middle row of the breadboard. The clocking notch of the chip (that little indentation or dimple on one end) should face the left of the board. Though this isn't *mandatory*, it's considered common practice among electronics folk.

3. **Insert the two fixed resistors, R2 and R3, into the board, following the schematic and the sample breadboard in Figure 14-2.**

   As noted in Chapter 4, the pins on integrated circuit chips are numbered counter-clockwise, starting at the clocking notch. So, if you're facing the breadboard with the 555 on it, and the clocking notch is on your left, Pin 1 is to the left of the clocking notch and Pins 2, 3, and 4 run in a column down the left side of the IC. On the right side of the IC, Pin 5 is opposite Pin 4, and Pins 6, 7, and 8 run up in a column (with Pin 8 opposite Pin 1).

4. **Insert the two capacitors, C1 and C2, into the board, following the schematic and the sample breadboard in Figure 14-2.**

5. **Solder wires to the potentiometer (R1) to connect it to the breadboard.**

   Use 22 gauge solid strand hookup wire. The color doesn't matter. Note that the potentiometer has three connections to it. One connection goes to pin 7 of the 555; the other two connections are joined (or "bridged") and attach to the V+ of the power supply.

6. **Connect the light-emitting diode as shown in the schematic and the sample breadboard.**

   You must observe proper orientation when inserting this component. You must connect the cathode of the LED to ground. Check the packaging that came with your LED to make sure you get it right. (If you don't, and you insert the LED backwards, nothing bad will happen, but the LED won't light. Simply remove the LED, and reinsert it, the other way around.)

7. **Use 22 gauge single strand wire, preferably already pre-cut and trimmed for use with a solderless breadboard, to finish making the connections.**

   Folks commonly refer to these wires as jumpers; most of the circuits you build will have at least one or two. Use the sample breadboard in Figure 14-2 as a guide to making these connections.

8. ***Before* applying power, double-check your work. Verify all the proper connections by cross-checking your wiring against the schematic.**

9. **Finally, attach a 9-volt battery to the V+ and ground rows of the breadboard.**

   The V+ row is on the top, and the ground row is on the bottom. It's easier to use a 9 volt battery clip, which you can get at RadioShack and other electronics stores. It's a good idea to solder 22-guage solid hookup wire to the ends of the leads from the clip; this makes it easier to insert the wires into the solderless breadboard. Remember: the red lead from the battery clip is V+; the black lead is ground.

When you apply power to the circuit, the LED should flash. Rotate the R1 knob to change the speed of the flashing. If your circuit doesn't work, disconnect the 9-volt battery, and check the connections again.

Here are some common mistakes you should look for:

- ✔ You inserted the 555 IC backwards. This can damage the chip, so if this happens, you might want to try another 555.
- ✔ You inserted the LED backwards. Pull it out and reverse the leads.
- ✔ You didn't press the connection wires and component leads into the breadboard sockets firmly enough. Be sure that each wire fits snugly into the breadboard, so there are no loose connections.
- ✔ The component values are wrong. Double-check, just in case!
- ✔ The battery died. Try a new one.
- ✔ You wired the circuit wrong. Have a friend take a look. Fresh eyes can catch mistakes that you might not notice.

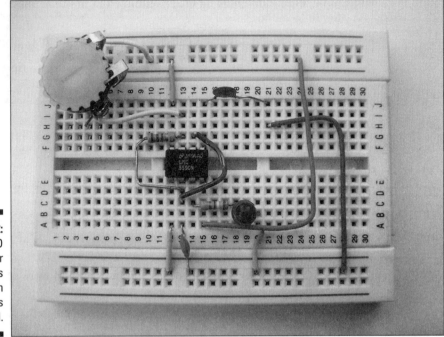

Figure 14-2:
An LED
flasher
with parts
mounted on
a solderless
board.

Its good electronics practice to build a circuit that's new to you on a breadboard first, as you often need to tweak a circuit to get it working just the way
you want it to. Once you have it working to your satisfaction on a breadboard,
then you can make the circuit permanent if you like. Just take your time, and
remember to double- and even triple-check your work. Don't worry — you'll
be a pro in no time, and building fairly complex circuits on your solderless
breadboard.

Running down the LED flasher parts

Here are the parts that you need to build the LED flasher circuit:

- **IC1:** LM555 Timer IC
- **R1:** 1 megohm potentiometer
- **R2:** 47 Kohm resistor
- **R3:** 330-ohm resistor

> ✔ **C1:** 1 µF tantalum (polarized) capacitor
>
> ✔ **C2:** 0.1-µF disc (non polarized) capacitor
>
> ✔ **LED:** Light-emitting diode (any color)

Putting the Squeeze on with Piezoelectricity

Not all electronic circuits require batteries, resistors, capacitors, transistors, or any of the other usual components that you find in an electronic circuit. This project generates its own electricity and you end up with a light drum consisting of a neon light that glows when you tap on a piezo disc. It serves as a great demonstration of something called piezoelectrics.

Piezo — what?

The term *piezo* comes from a Greek word meaning to press or squeeze. Many years ago, folks with too much time on their hands found that you can generate electricity when you press certain kinds of crystals really hard. Lo and behold, these same crystals change shape — though only slightly — when you apply electricity to them. It turns out that this find was an important discovery because we use piezoelectricity in tons of everyday gadgets, such as quartz watches, alarm buzzers, barbecue grill starters, and scads of other devices.

Experimenting with piezoelectricity

A simple and fun way to experiment with piezoelectricity is to get a bare piezo disc. You can find these discs at most electronics stores and also online. You can get them very cheaply; usually a dollar or less apiece.

Get a disc with the two wires already soldered onto it. Some discs only have one wire; these discs work just fine, too. You can clip a wire to the edge of the disc's metal for the ground connection.

Figure 14-3 shows a demonstrator circuit with one disc and one neon bulb. (Try RadioShack or other electronics stores to get the neon bulb.) Neon bulbs are special in that they don't light up unless you feed them at least 90 volts. That's a lot of juice! But the piezo disc easily generates this much voltage.

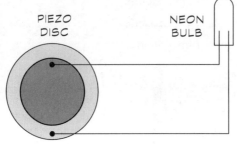

Figure 14-3:
Try this
simple
circuit to
demon-
strate the
properties
of piezo-
electricity.

PIEZO DISC POWERING A NEON BULB

To build the circuit in Figure 14-3, follow these steps:

1. **Place the disc on an insulated surface.**

   A wooden or plastic table surface works fine, but don't use a surface made of metal.

2. **Connect the disc and neon bulb together by using two alligator test leads, as shown in Figure 14-4.**

   Place one test lead from the red wire of the disc to one connection of the neon lamp (it doesn't matter which connection). The other test lead goes from the black wire of the disc to the other connection of the neon lamp.

3. **Place the disc flat on the table.**

4. **With the plastic end of a screwdriver, rap very hard on the disc.**

   Each time you rap the disc, the neon bulb flickers.

Avoid touching the two wires that come from the disc. Although the shock you get isn't dangerous, it definitely won't feel very good!

Need some ideas for how to use this concoction to impress your family and friends? How about building a light drum?

Follow these steps to build your very own light drum and dazzle your loved ones:

1. **String up a whole slew of discs and bulbs in a row.**

2. **Tape or glue these disc-bulb combos to a plastic base.**

3. **Get a pair of drumsticks, turn down the lights, and tap on the discs in time with your favorite mood music.**

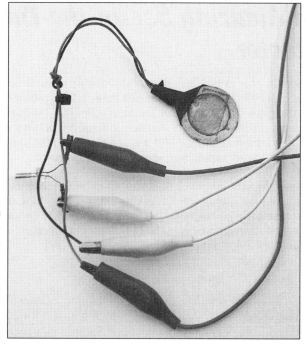

Figure 14-4:
Connect the disc and neon lamp, using alligator clips.

Gathering parts for the piezoelectricity circuit

For the circuit that demonstrates piezoelectricity, you need these very few parts:

- ✔ A bare piezo disc (the type that you use in a buzzer, preferably with two wires soldered on)

- ✔ Neon bulb

- ✔ Two alligator clips

- ✔ Something to whack the disc with, such as a screwdriver or drumsticks (not a baseball bat)

Building the Amazing See-in-the-Dark Infrared Detector

Did you ever want to see in the dark like a cat? Now you can, by building this simple infrared detector. The circuit uses just three parts (plus a battery). You can make the circuit a little fancier by adding an SPST (single-pole single-throw) toggle switch between the + (positive) side of the battery and the phototransistor to turn the detector on and off; or you can go the simple route and just unplug the battery when you aren't using the detector.

Figure 14-5 shows the schematic for the infrared detector. Be sure to use a phototransistor, and not a photodiode, in this circuit. They look the same on the outside, so check the packaging. Also, be sure to get the proper orientation for both the phototransistor and the LED. If you hook up either one backward, the circuit fails.

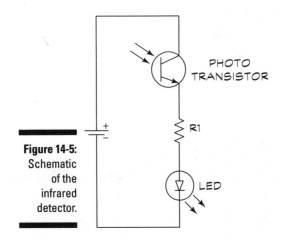

Figure 14-5:
Schematic
of the
infrared
detector.

Chasing down infrared light

Using the infrared detector, you can test for infrared light from a number of sources. Here are just two ideas to try:

> ✓ **Getting to the bottom of a remote control dilemma:** Because remote controls use invisible infrared light, you have a hard time figuring out what's wrong when they stop working. Does the remote have a problem,

or should you blame your TV or other appliance? To test the remote control, place it up against the infrared phototransistor. Press any button on the remote; if the LED on your project flashes, you know that you have a working remote.

✔ **Counter-surveillance:** Check to see if somebody's hidden camera is in your room. These days, covert cameras (such as the one in Figure 14-6) can see in the dark by using a built-in bright infrared light source. You can use the infrared detector circuit to find these sources, even if you can't see them yourself. Turn off the lights and scan the room by holding the detector in your hand and moving it around the room. If the LED brightens, even though you don't see a light source, you may have just found the infrared light coming from a hidden camera!

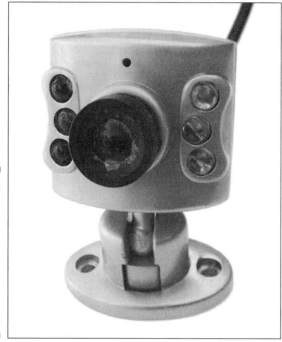

Figure 14-6:
This miniature camera can see in the dark, thanks to its six infrared-emitting diodes.

Although the infrared phototransistor is most sensitive to infrared light, it also responds to visible light. For best results, use the infrared detector in a dimly lit room. Sunlight, and direct light from desk lamps and other sources, can influence the readings.

Detecting parts for the infrared detector

Short and sweet, here's the list of what you need to build this project:

- ✔ **Q1:** Infrared phototransistor (our sample circuit uses a RadioShack 276-0145, but almost any phototransistor should work fine)
- ✔ **R1:** 330-ohm resistor
- ✔ **LED:** Light-emitting diode (any color)

Cheese It! It's the Cops!!

Unfortunately, you can't arrest any bad guys when you set off the warbling siren that you build in this project. But it sounds cool, and you can use it as an alarm to notify you if somebody's getting at your secret stash: Baseball cards, vintage Frank Sinatra records, your signed copy of *Mister Spock's Music from Outer Space* record, or whatever.

How your warbler works

This circuit (see Figure 14-7) uses two 555 timer chips. You rig both chips to act as astable multivibrators; that is, they constantly change their output from low to high to low to high . . . over and over again. The two timers run at different frequencies. The timer chip on the right in the figure produces an audible tone. If you connect a speaker directly to the output of this timer, you hear a steady, medium-pitch sound.

The output of the 555 chip on the left, which produces a slower rising and falling tone, connects into pin 5 of the 555 chip on the right in the figure. You connect the speaker to the output of the 555 chip on the right.

Adjust the two potentiometers, R2 and R4, to change the pitch and speed of the siren. You can produce all sorts of siren and other weird sound effects by adjusting these two potentiometers. You can operate this circuit at any voltage between 5 and about 15 volts.

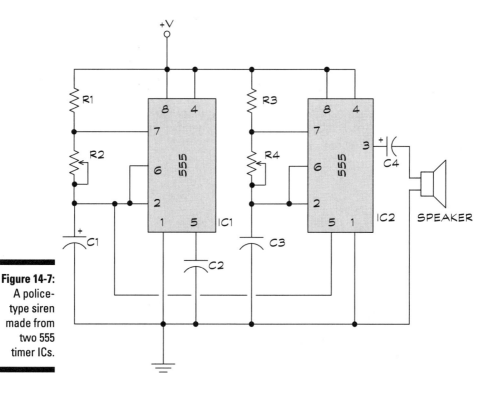

Figure 14-7:
A police-
type siren
made from
two 555
timer ICs.

Scoping out the 555 siren parts list

To start alarming your friends, gather these parts together to build the circuit:

- **IC1, IC2:** 555 Timer IC
- **R1, R3:** 2.2K ohm resistor
- **R2:** 50 Kohm potentiometer
- **R4:** 100 Kohm potentiometer
- **C1:** 47-μF electrolytic (polarized) capacitor
- **C2:** 0.01-μF disc (non-polarized) capacitor
- **C3:** 0.1-μF disc (non-polarized) capacitor
- **C4:** 1-μF electrolytic or tantalum (polarized) capacitor
- **Speaker:** 8-ohm, 1-watt speaker

Get Lost . . . or Found, with the Electronic Compass

Discover where in the world you are with this very cool electronic compass! This magnetic compass uses the same technology that manufacturers build into many cars to show you your direction electronically. Four LEDs light up to show you the four cardinal points on the map: N (north), S (south), E (east), and W (west). The circuit illuminates adjacent LEDs to show the in-between directions, SW, SE, NW, and NE.

Peeking under the compass hood

At the heart of this project is a special compass module, the Dinsmore 1490. This module isn't a common, everyday part. You have to special order it, but you can have a lot of fun with the project, making it worth the $13 to $15 that you pay for the compass module. Check out the manufacturer's representative at `www.robsonco.com` for the compass module, and don't forget to try other possible sources by doing a Google or Yahoo! search. Try the search phrase "dinsmore compass."

The 1490 compass module is about the size of a small thimble. The bottom of the sensor has a series of 12 tiny pins, as you can see in the pinout drawing in Figure 14-8. The pins are arranged in groups of four and consist of the following connection types:

✔ Power

✔ Ground

✔ Output (or signal)

You can see the schematic for the electronic compass in Figure 14-9. By doing some careful soldering, you can build a nice portable, electronic compass that you can take anywhere. Put it in a small enclosure, with the LEDs arranged in typical clockwise N, E, S, W circular orientation. You can buy enclosures at RadioShack and other electronics stores. They come in a variety of sizes, starting from about two inches square. Select an enclosure large enough to contain the circuit board and batteries.

You can power the compass by using a 9-volt battery. Add a switch from the + (positive) terminal of the battery to turn the unit on or off, or simply remove the battery from its clip to cut the juice and turn off your compass.

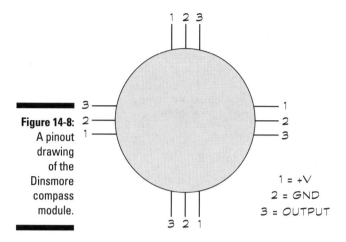

Figure 14-8:
A pinout
drawing
of the
Dinsmore
compass
module.

1 = +V
2 = GND
3 = OUTPUT

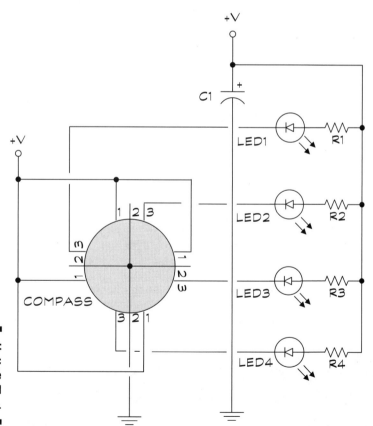

Figure 14-9:
Schematic
of the
handheld
compass.

Checking your electronic compass parts

To point you in the right direction, here are the parts that you need to gather to build your compass:

- ✔ **COMPASS:** Dinsmore 1490 magnetic electronic compass (see the section "Peeking under the compass hood" earlier in this chapter)
- ✔ **R1-R4:** 1 Kohm
- ✔ **C1:** 10-μF electrolytic (polarized) capacitor
- ✔ **LED1-LED4:** Light-emitting diode (any color)
- ✔ **MISC:** Project box, switch, battery clip (all optional)

When There's Light, You Hear This Noise . . .

Figure 14-10 shows you a schematic of a light alarm. The idea of this project is simple: if a light comes on, the alarm goes off. You build the alarm around an LM555 timer chip, which acts as a tone generator. When light hits the photoresistor, the change in resistance triggers transistor Q1. This response turns the 555 on, and it squeals its little heart out. You can adjust the sensitivity of the alarm by turning R1, which is a variable resistor (potentiometer).

Making your alarm work for you

Does it seem a little nuts to create an alarm that goes off whenever it senses light? Surprise! You can apply this handy light alarm in several practical ways. Here are just a few of 'em:

- ✔ Put the light alarm inside a pantry so that it goes off whenever someone raids the Oreo cookies. Keep your significant other out of your stash — or keep yourself on that diet! When the pantry door opens, light comes in and the alarm goes off.
- ✔ Do you have a complex electronics project in progress in the garage that you don't want anybody to disturb? Place the alarm inside the garage, near the door. If someone opens the garage door during the day, light comes through and the alarm sounds.
- ✔ Build your own electronic rooster that wakes you up at daybreak. (Who needs an alarm clock?)

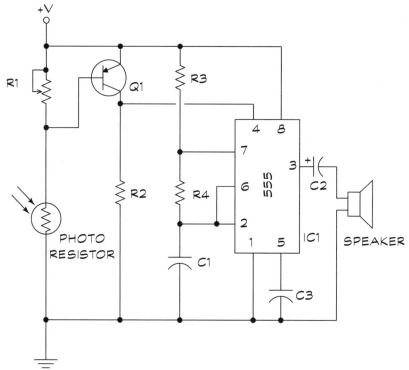

Figure 14-10:
Schematic
of a light
alarm.

Assembling a light alarm parts list

Here's the shopping list for the light alarm project:

- ✔ **IC1:** LM555 Timer IC
- ✔ **Q1:** 2N3906 PNP transistor
- ✔ **R1:** 100K potentiometer
- ✔ **R2:** 3.9 Kohm resistor
- ✔ **R3:** 10 Kohm resistor
- ✔ **R4:** 47 Kohm resistor
- ✔ **C1, C3:** 0.01-µF disc (non-polarized) capacitor
- ✔ **C2:** 1.0-µF electrolytic or tantalum (polarized) capacitor
- ✔ **Speaker:** 8-ohm, 0.5-watt speaker
- ✔ **Photoresistor:** experiment with different sizes; for example, a larger pho-toresistor will make the circuit a little more sensitive.

'Lil Amp, Big Sound

Give your electronics projects a big mouth with this little amplifier designed around parts that are inexpensive and easy-to-find at most electronics suppliers. LM386 power amplifier IC — this amp boosts the volume from microphones, tone generators, and many other signal sources.

The ins and outs of 'Lil Amp

Figure 14-11 shows the schematic for this project, which consists of just six parts, including the speaker. You can operate the amplifier at voltages between 5 and about 15 volts. A 9-volt battery does the trick.

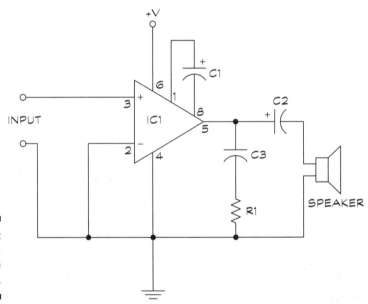

Figure 14-11:
Schematic
of the little
amplifier.

To use the amplifier, connect a signal source, such as a microphone, to pin 3 of the LM386. Be sure to also connect the ground of the signal source to the common ground of the amplifier circuit.

Depending on the source signal, you may find that you get better sound if you place a 0.1-μF to 10-μF capacitor between the source and pin 3 of the LM386. For smaller values (less than about 0.47-μF), use a disc capacitor; for larger values (1-μF or higher), use a tantalum capacitor. When you use a polarized capacitor, orient the + (positive) side of the component toward the signal source.

This little amp doesn't come with a volume control, and the sound quality can take you back to your days listening to the high school PA system. But this simple circuit puts out a whole lotta sound in a small and portable package.

Sounding the roll call for little amplifier's parts

Here's a rundown of the parts that you have to gather for this project:

- **IC1:** LM386 Amplifier
- **R1:** 10-ohm resistor
- **C1:** 10-μF electrolytic (polarized) capacitor
- **C2:** 220-μF electrolytic (polarized) capacitor
- **C3:** 0.047-μF disc (non-polarized) capacitor
- **Speaker:** 8-ohm, 0.5-watt speaker

The better the microphone and speaker, the better the sound!

Building the Handy-Dandy Water Tester

You may not be able to divine underground water with the water tester circuit in Figure 14-12, but it can help you check for moisture in plants or find water trapped under wall-to-wall carpet.

How the water tester works

The Handy-Dandy Water Tester is deceptively simple. It works under the principle of electrical conductivity of water (this is the principle that says you don't take a bath with a plugged-in toaster in your lap). The tester contains two small metal probes. When you place the probes in water, the conductivity of the water completes a circuit. This completed circuit drives current to a transistor. When the transistor turns on, it lights an LED. When the probes aren't in contact with water (or some other conductive body), your tester has a broken circuit, and the LED doesn't light up.

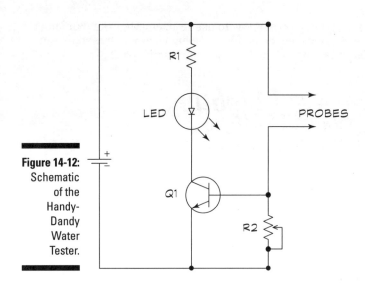

Figure 14-12:
Schematic
of the
Handy-
Dandy
Water
Tester.

You make the two probes with small nails, say 4d (four penny). Place the nails about a half-inch apart on a piece of plastic (but not wood or metal). The nails should be parallel to one another. File down the tips of the nails to make sharp points. These points help you drive the probes deep into the material that you're testing. For example, you can drive the probes into a carpet and pad to determine if water has seeped under the carpeting after a pipe in the next room bursts.

You can adjust the sensitivity of the tester by turning potentiometer R2. Start with the potentiometer in its middle position and turn one way or another, depending on the amount of moisture or water in the object that you're testing.

Power the water tester by using 5 to 12 volts. A 9-volt battery works well.

Gathering water tester parts

Go out and get the following parts to build your water tester project:

- ✔ **Q1:** 2N2222 NPN transistor
- ✔ **R1:** 470-ohm resistor
- ✔ **R2:** 50 Kohm potentiometer
- ✔ **LED:** Light-emitting diode (any color)
- ✔ **Probes:** Two small nails (4d, also called four penny)

Refer to the project description for how to use the nails because they aren't your standard electronics part!

Creating a Very Cool Lighting Effects Generator

If you were a fan of the *Knight Rider* television series that aired back in the '80s, you remember the sequential light chaser that the Kitt Car sported. You can easily build your own (light chaser setup, not car) in the garage in under an hour.

To build your own mesmerizing lighting effects generator (see the schematic for it in Figure 14-13), you need just two low-priced integrated circuits and a handful of inexpensive parts.

The circuit has two sections:

- **The brains:** An LM555 timer IC makes up the first section, on the left of the schematic. You wire this chip to function as an astable multivibrator (in fact, you make the same basic circuit as the LED flasher that we describe in the section "Creating Cool, Crazy, Blinky Lights" earlier in this chapter). The 555 produces a series of pulses; you determine the speed of the pulses by dialing potentiometer R1.

- **The body:** The second section, on the left of the schematic, contains a 4017 CMOS Decade Counter chip. The 4017 chip switches each of 10 LEDs on, in succession. The LEDs are switched when the 4017 receives a pulse from the 555. You wire the 4017 so that it repeats the 1-to-10 sequence over and over again, for as long as the circuit has power.

Arranging the LEDs

You can build the lighting effects generator on a solderless breadboard just to try it out. If you plan to make it into a permanent circuit, give some thought to the arrangement of the ten LEDs. For example, to achieve different lighting effects, you can try the following:

- **Put all the LEDs in a row, in sequence:** The lights chase each other up (or down) over and over again.

- **Put all the LEDs in a row, but alternate the sequence left and right:** Wire the LEDs so that the sequence starts from the outside and works its way inside.

✔ **Place the LEDs in a circle so that the LEDs sequence clockwise or counterclockwise:** This light pattern looks like a roulette wheel.

✔ **Arrange the LEDs in a heart shape:** You can use this arrangement to make a unique Valentine's Day present.

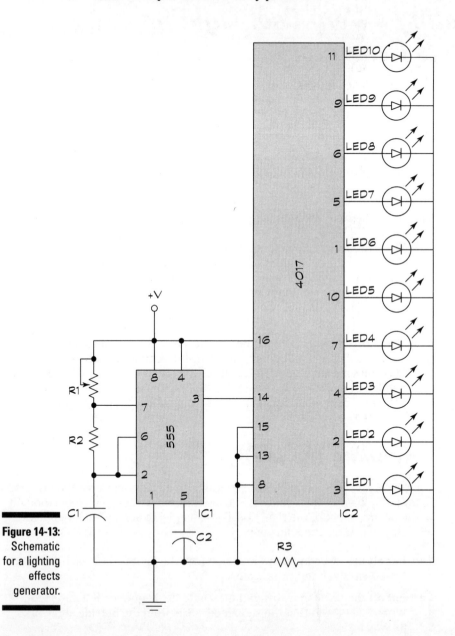

Figure 14-13:
Schematic for a lighting effects generator.

Going to the store for light chaser parts

To start chasing lights, you need the following parts:

- ✔ **IC1:** LM555 Timer IC
- ✔ **IC2:** 4017 CMOS Decade Counter IC
- ✔ **R1:** 1 megohm potentiometer
- ✔ **R2:** 47 Kohm resistor
- ✔ **R3:** 330ohm resistor
- ✔ **C1:** 0.47-µF disc (non polarized) capacitor
- ✔ **C2:** 0.1-µF disc (non polarized) capacitor
- ✔ **LED1-10:** Light-emitting diode (any color)

Chapter 15

Cool Robot Projects to Amaze Your Friends and Family

. . . .

In This Chapter

▶ Getting into the guts of a robot

▶ Preparing to build your very own 'bot

▶ Constructing Rover, a great beginner robot

▶ Giving Rover some smarts

▶ Adding motors, wheels, switches, and batteries for a complete 'bot

▶ Programming the Rover's BASIC Stamp 2 brain

▶ Diving farther into the wide world of robotics

. . . .

Make no mistake: Electronics is fun. But after you've built your 14th blinky light project, you yearn for more of a challenge. You look for bigger and better projects as you explore new facets of the electronic arts.

Robotics may be just what you're looking for. A robot is an amalgam of hardware, software, and electronics — all twisted together in a way that appears to bring life to a lump of plastic, metal, and silicon. Not long ago, building a robot meant toiling long hours in the garage and spending hundreds, if not thousands, of dollars.

Thanks to modern electronics, especially the microcontroller that lets you program a robot to perform all sorts of actions, you can build a robot for under $150. You get to decide what your robot does. You can have it seek out new life forms or explore the dark regions of your nephew's room. Or maybe you have a use in mind that no one has even thought of yet.

In this chapter, you build two robots. (Actually, you build one robot, but in two versions.) The first version is a simple 'bot with no brain. In the second version, you add a microcontroller, which you program to make your robot perform various tasks. With or without a brain, both 'bots give you the opportunity for a lot of fun.

Robots: The Big Picture

You're probably familiar with robots in the movies. These things walk, talk, and fend off alien armadas with their laser beam weapons. While there have been great strides in technology over the past several decades, today's robots aren't quite this fantastic. A robot that you build in your garage is more likely to be about the size of a cat, with less thinking capacity than a cockroach. This doesn't mean they aren't fun, though! On the contrary, playing around with small robots is a rewarding hobby, and they're getting cooler all the time.

The two projects we included in this chapter represent two distinct families of robots: human-controlled and autonomous:

- ✔ **You manipulate a human-controlled robot.** It's the person, not the robot, who does all the thinking. These work a lot like a remotely controlled racecar. The control may be wired or wireless. If you build one of these robots, it's an ideal way to get your feet wet because you can start out with a simple project and work your way up to more complex designs.

- ✔ **Autonomous robots think all on their own.** They have a small computer for a brain, and usually, one or more sensors so that they can detect their environment and respond to it. You program the robot's computer, and that controls all of the little critter's actions.

In this chapter you start out by constructing a human-controlled robot. The design is simple, consisting of a robot *base* that has two motors, two drive wheels, and a swivel caster that keeps it balanced. You control the robot by flipping two switches. Each switch controls one of the motors. You mount the switches on a little piece of wood or plastic, along with a few AA batteries. You connect the batteries and switches to the robot motors with long wires, so you can walk around the room while steering your robot.

Then, you will read all about how to build an autonomous robot. In this version, you do away with the switches and replace them with a BASIC Stamp 2 microcontroller to program the 'bot. The robot also uses a switch as a kind of bumper that tells the robot when it's run smack into something. You also discover how to program the microcontroller to make the robot steer in a new direction when the switch is bumped.

Rover the Robot parts list

Here are the parts that you need to build your robot (we tell you more about all of these shortly):

- Bottom deck, cut to size
- Top deck, cut to size
- 2 Tamiya worm gear motors (model #72004)
- 2 Tamiya Narrow Wheel sets (model #70145)
- 1¼-inch swivel caster
- 2 6-32 by ½-inch machine screws
- 2 6-32 nuts (for caster)
- 4 risers, constructed with standoffs or 6-32 machine screws
- 2 DPDT center-off toggle switches (Get toggle variety switches, with a center-off position. The switches should be spring-loaded momentary. That way, the switch handle will return to the center-off position when you release it.)
- 4-cell AA battery holder
- Small wooden or plastic board (about 4"x4" is fine) for mounting the switches and battery.
- 20 to 25 feet of flexible lamp (also called zip) cord
- Solder
- Electrical tape

Here are some notes to help you find these materials and parts:

- See the section "Gathering Your Materials" later in this chapter for some suggestions on the materials that you may want to use to build the top and bottom decks.
- You can purchase the Tamiya motors and wheels from Tower Hobbies (www.towerhobbies.com), as well as many other local and online hobby retailers. See the Appendix for a more complete list of sources.
- You can find the 1¼-inch caster at Lowe's and many other home improvement stores.
- Look for the DPDT switches, battery holder, and electrical items at RadioShack, or most electronics supply stores.

The following sections give you more details about all of these parts and materials and how you use them to build your robot.

The bits and pieces of a 'bot

You don't have to make your robot's body elaborate to make it good. You can make a simple and sturdy body by using common tools and readily available materials. Simple design choices can save you headaches. You can build a square-shaped body more easily than a circular one because the square needs only straight cuts. Cutting robot bodies from in-stock sizes of materials saves you money, too.

You can also decide whether you want to be more or less meticulous about how you construct your robot's body. For a lot of folks, building the mechanical body of a robot is akin to getting a root canal. They don't like all the cutting and sawing and drilling. So, they pull out the duct tape and invoke the physics of stickum. Although these construction techniques have their place, a sturdy and permanent body gives you a less hassle-prone robot, and things don't come off when they shouldn't.

Introducing Rover the Robot

The Rover, which we talk about in this chapter, is a fairly simple robot that gives you a perfect intro to robot building if you're just starting out. You can conveniently use the body of Rover for both projects in this chapter:

- ✔ In the basic Rover, you use two small DC gear motors to control its movement by using a pair of wired switches. (Gear motors are like regular motors, except they also include a set of gears that make the motor more useful for propelling things like small robots.) You can drive Rover through the house by flipping the switches up and down. It's a lot more fun than it sounds, especially if you have a cat or dog to chase around. (Don't worry — no animals will be harmed during this project.)

- ✔ The advanced Rover uses a microcontroller, specifically the BASIC Stamp 2, to build a self-contained, autonomous robot. With this version, you program the microcontroller for what you want Rover to do. This smart Rover uses specialized motors that you have to take apart and modify. You can read about how to do that in the section "Modifying the R/C servo motors" later in this chapter.

Preparing to Build the 'Bot

Before you can drill your first hole or fasten your first nut, you need to lay out the design of your robot, acquire all the materials, and sort out all the parts you're going to use.

First, get yourself a template

Although you use a circuit schematic for simpler projects, such as those in Chapter 14, you graduate to something called a cutting and drilling template for making a robot. This template serves as the layout for your robot. Draw the layout to scale — that is, the same size and shape that you want the finished pieces to be — on a piece of paper.

Figure 15-1 shows the cutting and drilling template for Rover, a two-deck tabletop robot. The dimensions used in this template are measured in inches.

The template includes the two body pieces, which we call decks, like the decks of a ship. There's a bottom deck and a top deck. You attach the motors and wheels to the bottom deck. The top deck is left free for future enhancements (such as adding the microcontroller brains, detailed later in this chapter).

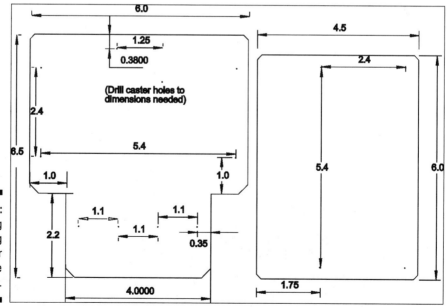

Figure 15-1: A cutting and drilling template for Rover the Robot.

The two decks are attached together using *risers*, which are long pieces of metal that are threaded to attach to machine screws. Read more about risers in the section "Getting to know the pieces," later in the chapter.

Gathering your materials

You need only rudimentary construction skills to build Rover using a variety of materials for the deck pieces. These are the easiest to work with:

- **¼-inch hardwood plywood:** A good choice is 5-ply "aircraft plywood," available at any hobby store.

- **¼-inch rigid expanded PVC sheet:** Commonly referred to as PVCX, Sintra, or Komatex.

- **⅛-inch acrylic plastic:** You can buy this plastic at most plastics specialty stores (you can find these by looking up "Plastics" in the Yellow Pages) and at many home improvement stores.

Our favorite construction material for a robot body is rigid expanded PVC because it's strong but lightweight, relatively cheap, and easy to cut and drill. You can sand it like wood, and, in fact, residential and commercial builders often use it as a wood substitute. Rigid expanded PVC is great stuff, but hardware and home improvement stores don't stock it. Look for it at specialty plastics outlets and sign-making shops. We provide some mail-order sources of small pieces in the Appendix.

Our least favorite construction material is acrylic plastic, for a number of reasons:

- If you're not careful, acrylic can shatter when you're drilling or cutting it.

- Acrylic dulls tools rather quickly.

- Acrylic generates a ton of static electricity, which, as we point out in Chapter 2, can damage sensitive electronics components.

Although acrylic plastic may not be the ideal material, you can use it in a pinch if you have nothing else suitable around.

Getting to know the pieces

You construct Rover's body with two pieces: a bottom deck and a top deck. The bottom deck measures 6 by 6½ inches, and you will cut out wells in it for the tires. The top deck measures 6 by 4½ inches, and it gives you enough room to mount all kinds of electronics and other goodies on it.

Getting savvy about machine screw sizes

In the United States and a few other places in the world, folks express machine screw size as two sets of numbers, such as 6-32. Here's what those numbers mean:

✔ **The first number:** This number represents the diameter of the screw. The smaller this number, the smaller the screw. (Screws that are ¼-inch or larger in diameter drop the number and just use the actual size, such as ⅜- or ⁵⁄₁₆-inch.)

✔ **The second number:** This number represents the threads per inch of the screw.

So, a 6-32 machine screw has a diameter defined as #6 (you don't have to worry about what that translates to, just get the one your project calls for or the one that fits), with 32 threads per inch. Other common screw sizes are 4-40, 8-32, and 10-24. In addition to the screw size, you need to know the length of the screw, such as ½- or ¾-inch, as well as the type of head on the screw. A screw you often use in building small robots has a round head.

You use a similar numbering scheme with metric screws, but the sizes and threads are expressed in millimeters rather than inches.

The drive motors for Rover, which propel the robot across the floor, are Tamiya worm gear motors (model number 72004). You can buy these parts at Tower Hobbies and many other online resources; check the Appendix for a list of some sources you can try. You buy the motor as a kit; you build it into a compact, self-contained housing. The motor comes with a shaft that you can attach to numerous styles of wheels.

Four risers separate the two deck pieces. You can use standoffs for the risers. *Standoffs* are lengths of metal that have threads on either end to accept common sizes of machine screws (these threaded ends are referred to as 'female'). The longer the standoff, the greater the distance between the two decks. You can get standoffs at electronics supply stores. Or, you can use 2- or 1½-inch long 6-32 machine screws that you can get at most hardware or home improvement stores.

Keep these points in mind when choosing what to use as risers:

✔ When using standoffs as risers, the minimum length that you can use is ½-inch; 1- or 1½-inch standoffs work even better.

✔ If you're using machine screws as risers, remember that the length of the screw must accommodate whatever distance you want between the deck pieces, plus the thickness of the deck piece material, *plus* the thickness of the retaining nut. For example, if you want 1 inch of clearance, and you're using ¼-inch thick wood or PVC, then the machine screws must be at least 1¾ inches. The remaining 1/4" is just enough for you to fasten a nut at the end of the screw.

Building the Body of the 'Bot

Now that you have your building plan and materials in hand, you're ready to actually start construction. So put on your hard hat, and read on! (Oh, and don't forget to grab safety glasses to protect your eyes.)

Cutting and drilling the pieces of a robot body

The first step in the building process is to use the robot layout from Figure 15-1 to drill holes for mounting parts, and then cut out the deck pieces.

Follow these simple steps to make the body pieces for your Rover robot:

1. **Lay out the holes and cutting dimensions from the template in Figure 15-1 directly on the wood or plastic material. Or better yet, draw them first on a piece of paper, then tape the paper over the wood or plastic material you are using for the decks.**

2. **When all looks right, drill the holes for Rover using a ⅛-inch drill bit.**

   You can use a hand drill (manual or motorized), but a drill press helps you to make more accurate holes. The distance and alignment of the four holes that you use to mount the two motors are the most critical.

3. **After you finish drilling, cut the pieces to size.**

   You can use a hacksaw (see Figure 15-2), coping saw, jigsaw, band saw, or scroll saw — whatever you happen to have in your shop. We prefer the scroll saw because it provides more control.

4. **Sand down the corners of the pieces to produce a beveled edge. This removes the sharp angles at each corner.**

   Use a motorized sander, such as the one that you see in Figure 15-3, to remove the sharp corners. But if you don't have this tool, use a sandpaper block with 60- or 80-grit paper and a little elbow grease to get the job done.

When you saw pieces without the benefit of a straight edge, you should cut a little outside the line that you marked onto the paper layout or material and then clean up any irregularities with a file or sandpaper block.

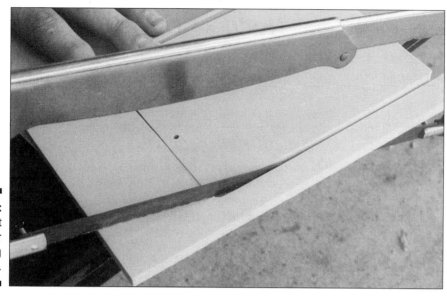

Figure 15-2:
You can cut
the Rover
pieces using
a hacksaw.

Figure 15-3:
Sand down
the corners
of the robot
decks so
they are
blunted.

Assembling and mounting the motors

Set aside the body pieces for now and turn to assembling the two gear motors. You put these motors together by following the instructions that come with your motor kits. Use a #1 Phillips screwdriver to assemble the motors; note that the screwdriver doesn't come in the kit, so you'll need to buy one if you don't already have it. You also need a small hex key wrench. You'll be happy to hear that the hex key wrench does come as a part of your motor kit.

Figure 15-4 shows how the gear motors should look when you've assembled them.

Attach the two motors to the bottom deck using some 6-32 screws and nuts. Refer to Figure 15-5, which shows you how to line up the motors with the holes in the bottom deck of your robot.

Figure 15-4:
A Tamiya gear motor, assembled and ready to go.

Figure 15-5:
The motors attach to the bottom deck of the Rover by using small screws.

Doing a wheelie

With the motors attached, it's time to secure the wheels in place. Putting the wheels for your little robot buddy together involves two steps — mounting an axle to the gear motor, and then attaching the wheels to that axle.

You get two styles of metal axle in the gear motor kit, one with and one without a hole on each end for a roll pin. (A *roll pin* is a small stick of metal that comes with the gear motor kit. It's tiny, so be careful not to lose it!) You use the axle that has the holes to build your robot. You can throw the other axle in your junk bin, ready for another project.

You have several options for the wheel, but our favorite is the Narrow Tire set (model number 70145), which comes with two 58-millimeter diameter rubber tires (about 2½ inches). Tamiya has included a hub that fits over the roll pin. You hold the wheel in place using a small hex nut, which also comes with the tire set. The wheels come as their own kit; follow their instructions to assemble them. Then, construct the second motor/wheel as a mirror image to the first.

When you use narrow tires, the motor shaft is a tad too long to fit in the space between each motor. You can easily fix this problem by cutting a small slit in the notch near the end of the motor shaft (opposite the end with the hole) and then breaking the end of the shaft off, using heavy-duty pliers.

Attach the two motors to the robot, using 4-40 by ½-inch machine screws and 4-40 nuts. Refer to Figure 15-6 to see how the motors and wheels should look after you attach them to the robot.

Mounting the caster

A swivel caster supports the base of the Rover robot, and is located on the opposite end from the gear motors and wheels. A standard 1¼-inch ball-bearing caster, which you can buy at Lowe's and other home improvement stores, works fine. Don't use a larger or heavier duty caster because these big boys weigh the robot down and don't swivel easily.

Figure 15-6:
The motors and wheels attached to our pal Rover.

You attach the caster by using 6-32 by ½-inch machine screws and nuts, as you can see in Figure 15-7. The drilling holes in the layout (see Figure 15-1) differ in size and spacing depending on the model of caster that you use. Your best bet is to trace the holes that you need to drill by using the caster base plate as a guide.

Adding the second deck

What you have at this point may look sort of like an open-faced sandwich robot. The next step, which completes the robot sandwich, is to add the top deck. You can add the second deck by using your choice of risers.

You can see the robot you've built so far, which still doesn't have any electronics, in Figure 15-8.

Figure 15-7:
Mount the caster to Rover by using 6-32 screws and nuts.

Figure 15-8:
The finished
(but not
smart)
Rover, with
wheels and
a second
deck.

Control switches

If you've gone through the preceding sections in this chapter, you've almost completed your Rover assembly. One last finishing touch, and you can start playing with it. Before you can play, Rover needs a power source in the form of a battery and a couple of switches so that you can control the operation of the motors.

Take a look at Figure 15-9. This diagram shows you how to hook up a battery to two double-pole double-throw toggle switches. This diagram also shows you how to wire the switches so that flipping a switch forward powers the motors one way and flipping the switch backward reverses the direction of the motor.

Follow these steps to hook up the battery to switches:

1. **Solder an 8 to 10 foot length of lamp cord wire from the center terminals of each switch to the left and right motor of the Rover.**

2. **Solder the leads from a four-cell AA battery holder to the switches.**

   The red wire and the black wire from the battery holder should connect to the terminals on both switches, as shown in Figure 15-9.

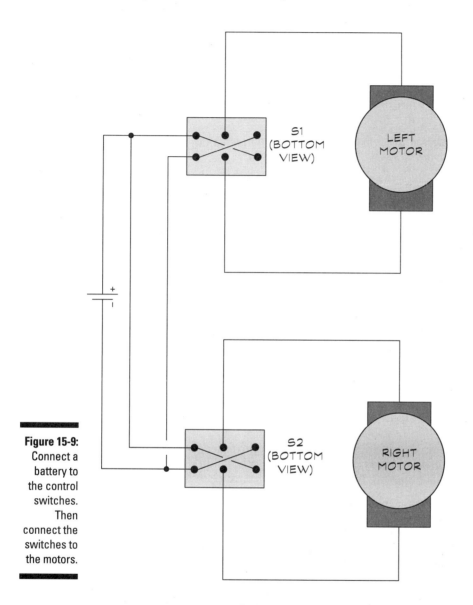

Figure 15-9:
Connect a
battery to
the control
switches.
Then
connect the
switches to
the motors.

3. **Solder the remaining "jumpers" between the switch terminals, as shown in the figure.**

   These jumpers form an X shape. When you wire them in this way, the switches reverse the polarity of the voltage from the battery as you toggle them from one position to the other.

4. **Drill two 3/8" (or so) holes in a piece of 4" x 4" x 1/4" plastic or wood.**

   The holes should be near the top of the plastic or wood piece, to allow room for the battery holder, which you'll place underneath. The exact size of hole depends on the mounting requirements of the switches you're using. The holes should be just large enough for the shaft of the switch to fit through.

5. **Use the retaining nuts that come with the switches to secure the switches to the plastic or wood.**

6. **Place the switches in their center/off position.**

7. **Put four AA batteries in the battery holder.**

8. **Use tape (electrical tape or duct tape works best) to secure the battery holder to the piece of plastic or wood.**

   Tuck some of the wire leading from the switches to the robot inside the tape. This acts as a kind of strain relief, and prevents the robot from pulling its wires out of the switches.

For best results, solder all connections and then use insulating electrical tape to cover any exposed wires. Wrap tape several times around each wire where it connects to each motor. This tape layer helps to keep the wires from pulling out.

To control the Rover, use your thumbs to push the switches back and forth. Release the switch, and it returns to its center, which is the off position. When the switch is off, the motor that the switch is connected to stops.

Because you mount the motors in mirror image, to go forward (or backward), one motor turns clockwise and the other motor turns counter-clockwise. Rotate the switches on the wood or plastic so that you can press both switches forward with one motion to move the robot forward and press both switches backward to reverse direction.

Driving Miss Rover

After going through the setup in the preceding sections in this chapter, you can take your robot out for a spin. You steer Rover like this:

- ✔ Make your robot turn by pushing one switch forward and the other switch back.

- ✔ Press the switches to get the robot to move forward, go backward, or make turns.

The arrows on the wheels of the robot representation in Figure 15-10 show you the direction that you flip the switches to achieve the motion that you want. In this figure there are only two wheels; we've added a caster at the front of our robot for balance, but the two rear wheels drive the robot and steer it, just as the two wheels in Figure 15-10 do.

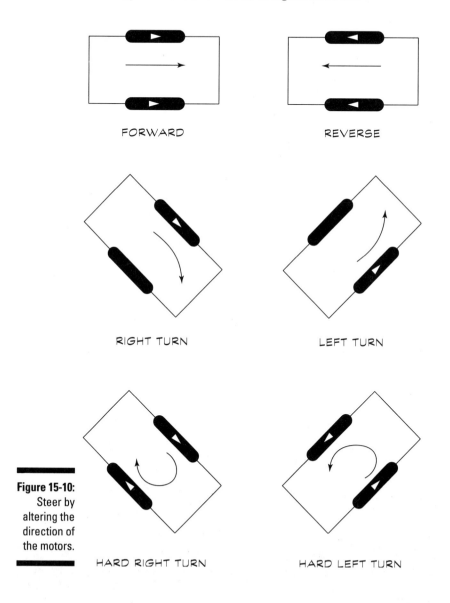

FORWARD   REVERSE

RIGHT TURN   LEFT TURN

Figure 15-10:
Steer by
altering the
direction of
the motors.

HARD RIGHT TURN   HARD LEFT TURN

Giving Rover Some Smarts

You control the basic Rover, which you can read about in the previous sections, by wire. You can have some fun driving Rover around the den, but some folks would say you don't have a true "robot" because it doesn't think on its own. Well, you can add a brain to Rover with way less effort than you may think. All it takes is a few changes to the motors so that they can accept electronic commands, the addition of a sensor or two to tell the robot that it's run up against an obstacle, and a microcontroller to tell the robot what to do when it does hit a snag (or a chair, or a wall).

To program your robot, you need to use a microcontroller. At the heart of the Smart Rover is a BASIC Stamp 2 microcontroller. We introduce these puppies in Chapter 13. If microcontrollers are a new concept to you, you can go and leaf through that chapter to get a feel for them.

For this project we assume that you have plugged a BASIC Stamp 2 into a BOE development board. (For more about these two items, read the next section or head back to Chapter 13.)

Mulling over microcontrollers

If you feel confident enough in your microcontroller knowledge to go on, we quickly recap:

- A microcontroller is a computer in the form of a chip that you connect to your robot through I/O (input/output) ports. By downloading programs from the computer to the robot, you can literally control the robot's actions.

- You program the microcontroller just as you program a desktop computer. You tell the microcontroller what to do, and it tells the robot.

- You write this program on your PC and then download it to the microcontroller, usually through a serial or USB cable.

- After you download the program into the microcontroller, it's stored in non-volatile memory, and here the program stays until you replace it with another program. The program remains in memory, even when you turn the power off and remove the batteries.

In Chapter 13, we discuss a nifty little microcontroller named the BASIC Stamp 2, from Parallax. This microcontroller was designed with the beginning robot builder in mind. The BASIC Stamp 2 is a 24-pin integrated circuit. You can use this integrated circuit right out of the box in your projects, or stick it into a development board, such as Parallax's Board of Education (BOE) to allow you to easily experiment with the BASIC Stamp. Chapter 13 also describes the BOE in more detail, so check it out to get more background.

DC motors out, R/C servo motors in

Before you get to program a microcontroller, you have to use different motors so that you can control them electronically. The basic Rover that we describe in the first project in this chapter uses two DC gear motors. These motors work ideally for switch control, allowing you to readily change the direction of the motor just by flipping a lever. But these kinds of motors need a bit more circuitry before you can operate them electronically. Specifically, they need an H-bridge, which does electronically pretty much what mechanical switches do in the basic Rover.

We promise not to make you buy or build an H-bridge; instead, use two inexpensive (about $10 to $12 each) servo motors that are designed for use in radio-controlled (R/C) gadgets such as model airplanes. The BASIC Stamp 2 can operate these motors directly, without any additional circuitry.

You can see an example R/C servo motor in Figure 15-11.

Figure 15-11:
A typical standard-size R/C servo motor.

Going inside a servo motor

So what's inside a servo motor? Servo motors that you use for radio-controlled planes and cars consist of some control electronics, a motor, a couple of gears, and a variable resistor (potentiometer).

Here's how all of these parts work together:

- The control electronics are there to receive a signal from a radio control receiver (or in our case, the BASIC Stamp microcontroller) to activate the motor.

- The motor turns. The motor runs pretty quickly, and your robot can't run that fast, so the action of a series of gears reduces the motor's speed.

- Connect the potentiometer to the final output gear. This gear protrudes outside of the servo motor and connects to a linkage, wheel, or whatever. As this output gear turns, so does the potentiometer. The position of the potentiometer tells the control electronics the position of the output gear.

Going shopping for servos

First things first — you need to go out and buy your servo motors. You can find more than a half-dozen major manufacturers of R/C servo motors, and each manufacturer offers a multitude of models. But when it comes to adapting a servo motor for robotics, only three servos stand out from the crowd as both affordable and easy to modify:

- Futaba S148
- Grand Wing Servo (GWS) S03 and S06
- Hitec HS-422

These servos share a common trait — they use a small retainer clip on the underside of the output gear to engage with the potentiometer. You can most easily modify this type of servo. When you remove the clip, the output gear no longer turns the potentiometer shaft. (You also need to clip with snippers or file off a molded-in ridge on the top of the output gear. This ridge serves as a physical stop to prevent the output gear from turning more than about 270 degrees.)

Making servos serviceable

Now you have your servo motors, but there's a fly in the ointment: The manufacturers designed the R/C servo motor to move back and forth about 90 degrees in each direction, and no more. To operate a robot, you need to convert the motor for continuous rotation so the wheels don't just stop at 90 degrees. Happily, this job isn't tough, as long as you choose the right kind of servo. (You can read more about how to modify an R/C servo motor in the following section.)

The benefit of modifying an R/C servo motor is that you don't need to monkey around with additional interface electronics to use the motor with a microcontroller, such as the BASIC Stamp 2. In addition, R/C servo motors are fairly low-priced, as motors go. These two facts make modifying servo motors well worth your while.

An unmodified servo allows precise positioning of the output gear. In a modified servo, you sever the link between the potentiometer and the output gear. The output gear then turns freely, without stopping.

After you modify a servo, you can simply plug it into sockets on the BOE, and it works like a charm. You need to write some programming code to operate the servo, but don't worry, we show you just what to do in the following sections.

Once you've done these modifications, the motors can now turn the wheels to steer Smart Rover (as shown in Figure 15-10) and move it around your living room.

Modifying the R/C servo motors

Here's what you need to modify any one of the servos that we recommend in the section "Going shopping for servos" earlier in this chapter:

- ✔ #0 Phillips screwdriver
- ✔ ⅛-inch or smaller flat-bladed screwdriver
- ✔ Nippy cutters, X-ACTO blade, or razor saw
- ✔ Small flat jeweler's file

After you modify a servo, you void its warranty, so be sure that you have a good 'un first. Test the servo for proper operation by plugging it into a microcontroller and sending it a command (see the section "Putting the program in place" later in this chapter for more about how to do this) before you modify it. Though it happens only rarely, a servo may fail right out of the box.

TIP

Throughout the following steps, take care to avoid wiping off too much of the lubricant that you use for the servo's internal gears. Otherwise the servo gears may not have enough lubrication for smooth operation. If you think you've lost too much of the lubricant, you can always add more just prior to re-assembling the motor. You can get servo grease at the same hobby store where you bought the servo motors.

Now, you're ready to make the actual modifications to your servo. We wrote the following steps for the Hitec HS-422 servo, but you can modify most other servos using the retaining clip design in much the same way:

1. **Use the Phillips screwdriver to remove the servo disc, if it attaches to the output gear/shaft.**

2. **Loosen the four casing screws from the bottom of the servo (see Figure 15-12).**

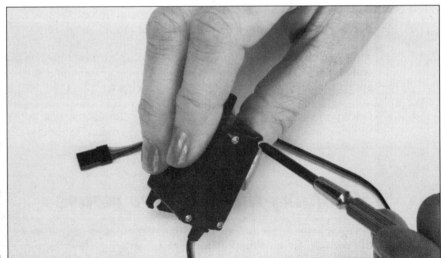

Figure 15-12:
Taking the servo apart.

3. **Remove the screws completely so that you can set the servo base down on the table while you're working inside it.**

   On a few servos, notably the GWS S03, you remove the case screws from the top, not the bottom.

4. **Remove the top portion of the servo and observe how all the gears are oriented so you can put things back the same way when you're done.**

   Look at Figure 15-13 for an example of what the innards of a servo look like.

Figure 15-13:
A taken-apart servo. Note the placement for all the gears.

5. **Remove the center gear, being careful not to unseat its metal shaft, and place the center gear aside.**

   On the Hitec HS-422, you can't easily remove the center gear without also lifting up the output gear, so carefully lift (and then replace) the output gear, if you need to.

6. **Remove the output gear.**

7. **Use a small pair of pliers to set the potentiometer at its center position, as you can see in Figure 15-14.**

8. **Remove the ridge on the top side of the output gear by using the nippy cutters (see Figure 15-15), an X-ACTO blade, or a razor saw.**

   Exercise caution! The harder the plastic, the more likely it is that the ridge will break suddenly and fly off. Wear eye protection. Always nip first on the long side to prevent the output gear from breaking. When using an X-ACTO blade or razor saw, observe the obvious precautions against cutting your fingers off. If you're using cutters, chip off small amounts of the ridge at a time, instead of trying to clip it off all at once. Otherwise, the pressure of the cutter can cause the output gear to break apart.

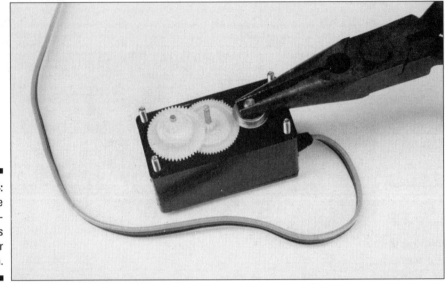

Figure 15-14:
Setting the potentio- meter to its center position.

9. **File down the small portion of the ridge that you're stuck with, no matter what cutting technique you use; do this filing with a small, flat file (see Figure 15-16).**

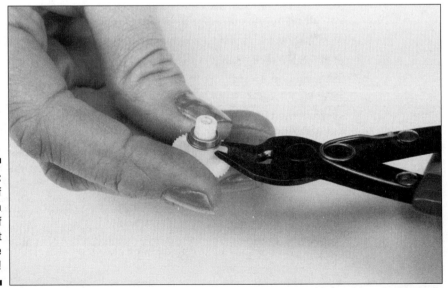

Figure 15-15:
Clipping off the ridge on the top of the output gear. Use caution!

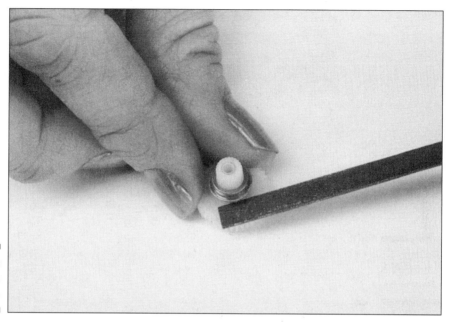

Figure 15-16:
Filing down
the ridge.

10. **Remove the metal retaining ring from the underside of the output gear (Figure 15-17), using the small-bladed screwdriver.**

    This ring holds the potentiometer shaft clip and supports the output gear.

11. **Use the small-bladed screwdriver again to remove the potentiometer shaft clip, as shown in Figure 15-18.**

12. **Place the metal retaining ring back into the output gear.**

13. **Replace the output gear on its seat, resting over the potentiometer.**

14. **Replace the middle gear and make sure that all gears mesh properly.**

15. **Add more grease at this point, if necessary.**

16. **Finally, put the top case back on and screw in the four case screws.**

Mounting the servos to the Rover

R/C servo motors have a screw flange that you can use to mount them permanently, but for the Smart Rover, just stick on some double-sided tape or Velcro to get the job done.

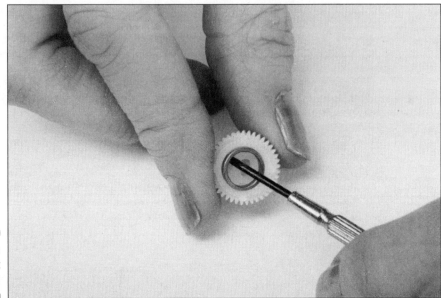

Figure 15-17:
Taking out
the retainer.

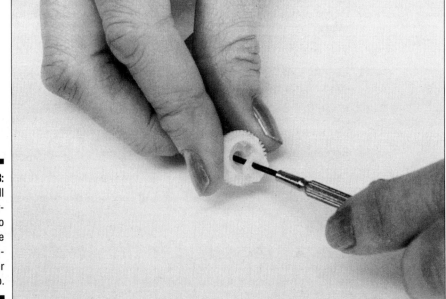

Figure 15-18:
Use a small
screw-
driver to
remove the
potentio-
meter
shaft clip.

We like to use a Velcro-like material called Dual Lock. 3M makes it, and you can find it at discount department stores, such as Target, and some hardware stores. Dual Lock works better at holding the parts of your robot together than Velcro because it doesn't permit as much side-to-side slippage.

Attach a piece of Dual Lock to the side of a servo. Stick a mating chunk of Dual Lock on the underside of Rover's bottom deck and squeeze the two pieces together to make a solid joint.

Figure 15-19 shows you what the servo looks like when you've mounted it on the Rover.

When you attach the servos to your Rover, make sure that you get both servos on straight. Otherwise, the robot may wobble around the room like a toddler learning to walk. Be sure to leave enough clearance for the wheels; otherwise the wheels may scrape against the robot.

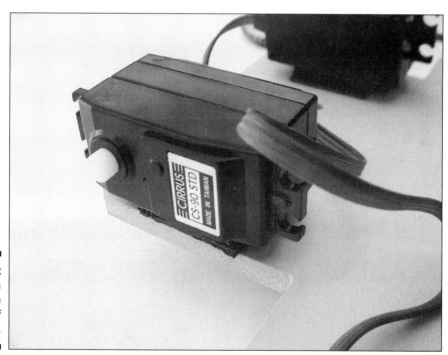

Figure 15-19:
Securing a servo to the bottom of Rover.

Putting Your Servos on a Roll with Wheels

The motors on Rover have one reason for being — to move the wheels and make the thing go. For the next step, after you attach the servo motors to the robot body (which we talk about in the preceding section), you have to attach wheels to the motors.

Several online and mail-order outfits, such as www.budgetrobotics.com and www.solarbotics.com, sell 2½-inch diameter wheels that you can attach directly to an R/C servo motor. (You can make your own wheels, but servo-ready wheels don't cost much, and are easy to use, so why bother?) Find some wheels that you like, and be sure to buy two. When you buy wheels, select the correct type for the servo that you're using. The mounting hub differs ever-so-slightly between Futaba and Hitec servos, so get two wheels of the same brand.

Servos made by Grand Wind (GWS) and Parallax use Futaba hubs. For these models, you have to purchase Futaba-style wheels.

Use a small screw inserted in the center hole of the wheel to attach the wheels to the output shaft of the servo; this screw comes with the servo. Figure 15-20 shows the Rover with wheels attached to the servo.

Figure 15-20:
Rover has wheels!

Sensing Things with a Bumper Car Switch

Rolling around the room all day is well and good, but eventually Rover smashes into the dining room table. That's where the bumper car switch comes in. The Smart Rover uses this small spring-operated leaf switch to help it figure out that it's run into something. The switch is a long bar located at the front of the robot, so that when the Rover hits something, the switch is triggered and the program running on the Rover's BASIC Stamp (more about this program in the section "Putting the program in place" later in this chapter) causes the robot to back up a short distance and then go the other way.

You can get an SPST spring-loaded leaf switch at almost any electronics parts store. You don't need to worry about size, as long as you have a switch big enough to mount to the front of the robot with double-sided tape, Velcro, or Dual Lock.

Figure 15-21 shows a diagram of a leaf switch with a three-inch length of 1/16" diameter brass piano wire soldered onto the leaf (but you can also just glue it on). The rod acts to extend the lever of the switch so that the Rover has a larger bumper contact area. You can buy piano wire at any hardware or hobby store.

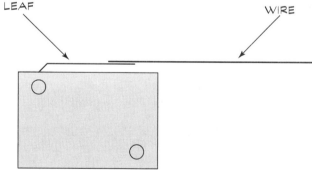

Figure 15-21: Diagram of a leaf switch with a piece of piano wire stuck to it.

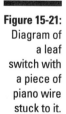

 Many leaf switches are SPDT. They have three terminals: common, normally open (NO), and normally closed (NC). These terminals work well for the Smart Rover. Just be sure to connect the common and NO (normally open) terminals and leave the NC terminal alone (we go into the procedure for this in the next section).

Figure 15-22 shows you where to mount the switch on the front of your Rover.

Figure 15-22:
Putting a
switch on
the front of
your robot.

Connecting Up to the Board of Education

You almost have a working smart 'bot! You're now ready to connect the servos, which you mount on the Rover (as we describe in the section "Mounting the servos to the Rover" earlier in this chapter), to the BASIC Stamp Board of Education. You can check out the overall connection scheme in Figure 15-23.

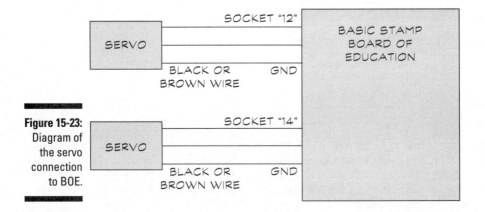

Figure 15-23:
Diagram of
the servo
connection
to BOE.

You use I/O ports 12 (labeled P12 in Figure 15-24) and 14 (labeled P14) to communicate between the BOE and the servos. Port 12 is located on pin 17 and Port 14 is located on pin 19 of the BASIC Stamp 2 chip.(Note that only pin numbers for the four corner pins are indicated in this figure.) Be sure to review the documentation that comes with your BOE for more information on the BASIC Stamp's I/O ports, pins, and other features.

You can access the I/O pins at a special jack on the BOE, as Figure 15-25 shows. The jack is designed to accept the connectors used on R/C servo motors. You just plug the connectors into the sockets marked 12 and 14, and you're all set. Easy as pie!

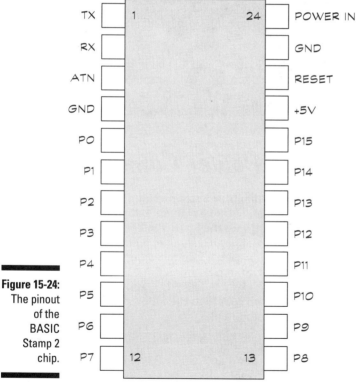

Figure 15-24:
The pinout of the BASIC Stamp 2 chip.

Figure 15-25:
Connections
to BOE.

Making Switch and Power Connections

Along with the switch you added in the earlier section, "Sensing things with a bumper car switch," you add an LED to Rover so that you get a visual alert when the switch is triggered. You wire the leaf switch to the board using the same connection that we detail in Chapter 13. Take a look at the section "Adding a Switch to the Mix" in Chapter 13 for more information about how to hook up a switch to the BASIC Stamp Board of Education.

The Smart Rover program that we go into in the following section also uses the LED indicator that we cover in Chapter 13.

Servo motors use up more current than a 9-volt battery provides. Use 4 AA batteries in a battery holder, rather than a single 9-volt battery. You use a special power plug with the BOE that you can buy at RadioShack or other electronics stores. Or, you can purchase a AA battery holder from Parallax with the proper power plug already attached; you can visit their Web page at www.parallax.com.

Don't get a battery that exceeds 6 volts, or you may damage the servo motors. Servo motors are designed to run at between 4.8 and 6 volts. Four rechargeable AA cells provide 4.8 volts; four alkaline AA cells provide 6 volts.

Making the Smart Rover Smart

Now the BASIC Stamp has to endow the Smart Rover with some brains. Check out the program listing below. It shows you how the program operates the two servos and reacts to the leaf switch:

✔ The program starts both servos, propelling the robot forward.

✔ If the robot hits something, the impact triggers the switch, and the program reverses one of the motors.

✔ The program reverses that motor for about three quarters of a second, which makes the robot spin around.

✔ The robot moves forward again until it, inevitably, hits something else.

Putting the program in place

In this section, we give you the program that you need to get your Rover rolling. Enter it into the BASIC Stamp editor, as we describe in Chapter 13, and run the program when you're done; this uploads it to your 'bot.

Don't forget that you need to connect your PC to the Board of Education through a serial or USB cable, depending on the BOE version you have. Read Chapter 13 for the complete picture.

```
'  {$STAMP BS2}
OUTPUT 0
btn    VAR    Byte        ' set up BUTTON variable
cnt    VAR    Byte        ' set of FOR/NEXT variable
loop:
  PULSOUT 12,1000          ' motor A
  PULSOUT 14,500           ' motor B
  PAUSE 15                 ' wait 15 milliseconds
  BUTTON 1,0,255,250,btn,0,noSwitch
  OUT0 = btn              ' turn LED on
  FOR cnt = 1 TO 50       ' count to 50 iterations
    PULSOUT 12,1000         ' motor A
    PULSOUT 14,1000         ' motor B
    PAUSE 15                ' wait 15 milliseconds
  NEXT
  OUT0 = 0                ' turn LED off
noSwitch: GOTO loop        ' repeat loop
```

Hmmm. You say your robot goes backward, rather than forward? You can fix that problem easily. Simply reverse the timing instructions in the program for motor A and motor B immediately following the loop: label:

```
PULSOUT 12,500          ' motor A
PULSOUT 14,1000         ' motor B
```

Looking at the program up-close

Taking a closer look at how the Smart Rover program works helps you become more programming savvy. Here's a blow-by-blow description of what each line does:

```
' {$STAMP BS2}
```

This line tells the BASIC Stamp editor that you're using BASIC Stamp 2.

```
OUTPUT 0
```

This line tells the BASIC Stamp to treat I/O port 0 as an output. The indicator LED connects to I/O port 0, and the program control turns the LED on and off.

```
btn    VAR    Byte
cnt    VAR    Byte
```

These lines set up the BASIC Stamp with two variables. These variables ensure that temporary data is stored until it's used again later in the program.

```
loop:
```

This line is the main loop of the program. It tells the BASIC Stamp to repeat the instructions from this point to the GOTO loop instruction at the bottom of the program. These instructions repeat over and over again. Get used to this command if you plan to do much 'bot building — almost all robot control programs have one.

```
PULSOUT 12,1000         ' motor A
PULSOUT 14,500          ' motor B
```

Pulses operate R/C servos. The length of the pulse determines the direction of travel. The PULSOUT programming statement sends a pulse of a specified duration to the indicated I/O port. For example, PULSOUT 12,1000 sends a 2,000-microsecond pulse to I/O port 12. (You specify the pulse duration in 2-microsecond increments: 1000 in the code equals 2000 microseconds.) But, hey, motor A pulses at 2000 microseconds, while motor B pulses for only 1000 microseconds. Why? Because you mount the motors in mirror image. To move the robot forward, one motor has to turn clockwise, and the other turns counter-clockwise.

```
PAUSE 15
```

Setting the exact center of your servos

When an unmodified R/C servo motor receives a series of pulses 1000 microseconds long, it moves as far as it can go in one direction. When it receives a series of pulses 2000 microseconds long, it moves as far as it can go in the other direction. As you may have guessed, when R/C servos receive a series of pulses 1500 microseconds long, the motors move to the center position. Makes sense, doesn't it?

When you modify your two servos, as we describe in the "Modifying the R/C servo motors" section of this chapter, you set the potentiometer to its center position. Well, there's physical center, and then there's electrical center. You only move the potentiometer to its center physical position, but not its electrical center position. You can make your Rover easier to control by setting the potentiometer to its electrical center. You do this Rover taming by running a short program and turning the potentiometer until all motor activity stops. Of course, you have to disassemble the servos again to reach the potentiometer.

Here's the program to use for this positioning (the program assumes that you have the servos connected to the Board of Education, as we discuss in the "Connecting Up to the Board of Education" section in this chapter):

```
' {$STAMP BS2}
loop:                    ' define start of
                           loop
    PULSOUT 12,750       ' motor A
    PULSOUT 14,750       ' motor B
    PAUSE 15             ' wait 15
                           milliseconds
    GOTO loop            ' repeat loop
```

This program is pretty simple. It sends out an endless stream of 1500-microsecond pulses to both servos. While it's sending these pulses, adjust the potentiometer until the motors stop.

Mind you, you don't absolutely need to do this step, but you may find it handy if you ever decide to delve deeper into the black art of robot building.

This line tells the BASIC Stamp to wait a brief period of time, specifically 15 milliseconds.

```
BUTTON 1,0,255,250,btn,0,noSwitch
```

The BUTTON statement tells the BASIC Stamp to check the state of the switch connected to I/O pin 1. The BUTTON statement requires a bunch of additional options, which you can find in the BASIC Stamp documentation that came with your BASIC Stamp kit. We discuss this programming statement more fully in Chapter 13.

```
OUT0 = btn
FOR cnt = 1 TO 50
    PULSOUT 12,1000
    PULSOUT 14,1000
    PAUSE 15
NEXT
```

These programming statements run if, and only if, the switch is triggered. The `OUT0 = btn` statement turns the LED on. The `FOR/NEXT` loop repeats the three statements that follow a total of 50 times. After the 50th iteration, the program continues. The pulsing causes one of the servos to reverse, so the Rover spins around and heads off in a new direction.

```
OUT0 = 0
noSwitch: GOTO loop
```

After the robot turns around, the LED turns off (as the first line above tells it to), and the main loop repeats itself (according to the command in the second line).

Where Can I Go from Here?

Obviously, you can discover much more about controlling robots using a BASIC Stamp or other microcontroller. If you're interested (and why wouldn't you be?!), check out the documentation that comes with your BASIC Stamp. That information includes several robot-related examples. Parallax, the makers of the BASIC Stamp, also offer several robot-programming tutorials and kits that you may find very handy.

And, lastly, don't forget the rich sea of the Internet. Although you can sometimes have trouble finding exactly what you want while swimming around out there, persistent use of search engines, such as Google and Yahoo!, help you dig up wonderful chestnuts of robot info when you search with terms such as "robot" or "robotics".

Part VI
The Part of Tens

The 5th Wave   By Rich Tennant

CRUDE HAM SHACK

In this part . . .

Since the dawn of Dummies time every *Dummies* book has a handful of "top ten" lists. These lists contain all sorts of useful chestnuts that, if nothing else, make for great reading while waiting for the dentist to call you in for your teeth cleaning. (And don't you wish these were "top twenty" lists, so you'd have an excuse to read longer!)

In this part, we offer the top ten additional testing tools for your electronics bench; ten great sources for electronics parts; and ten useful — but not overly boring — electronic formulas that you mathematically inclined folks will just love.

Chapter 16

Ten (Or So) Cool Electronics Testing Tool Tips

In This Chapter

▶ Using a logic pulser to inject test signals into circuits

▶ Checking the frequency of a signal

▶ Powering your gadgets with a variable power supply

▶ Generating waveforms with a function generator

▶ Sweeping signals with a sweep generator

▶ Checking inputs and outputs with a logic analyzer

▶ Viewing radio waves with a spectrum analyzer

▶ Injecting signals into an analog circuit

▶ Searching for static with a static meter

▶ Finding great deals on testing tools

*O*kay, so you're ready to graduate to the electronics big time. But you can't do it alone. You need a laboratory full of impressive-looking gear with blinky lights, bright knobs, and spinning dials. You're ready to go out and acquire some of the neat specialized test equipment that we describe in this chapter.

You don't absolutely, positively need these tools just to play around with some LEDs and resistors. A basic multimeter, and maybe a logic probe, are all you need for that. Consider acquiring the additional test gear in this chapter after you've gained some experience in electronics and want to graduate to bigger and better projects. Unless you're independently wealthy, just purchase test equipment as you need it.

Put a Pulse Here, Put a Pulse There

The logic pulser is a handy troubleshooting accessory for when you work with digital circuits. This handheld tool, which you can see in Figure 16-1, puts out a timed high or low digital pulse, letting you see the effect of the pulse on your digital circuit. (A *pulse* is simply a signal that alternates between high and low very rapidly.) Such a pulse might be used to trigger some portion of a circuit that is not otherwise working, for example — you can think of it as a way to "jump start" a cranky circuit. You can switch the pulser between one pulse and continuous pulsing. Normally, you'd use the pulser with a logic probe or an oscilloscope. (You can read about both logic probes and oscilloscopes in Chapter 10.)

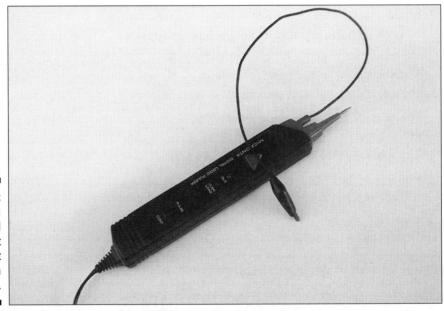

Figure 16-1:
Logic pulsers feed a short signal burst into a circuit.

Most pulsers get their power from the circuit that you're testing. You need to remember this fact because, with digital circuits, you generally don't want to present an input signal to a device that's greater than the supply voltage for the device. In other words, if a chip is powered by five volts, and you give it a 12-volt pulse, you ruin the chip.

Be sure that you don't pulse a line that has an output but no input. Some integrated circuits are sensitive to unloaded pulses at their output stages, and you can destroy the chip by applying the pulse improperly. (An unloaded pulse means that the current from the pulse has no way to safely drain to another part of the circuit. If the current is applied to an output of an integrated circuit, for example, that output *could* be damaged because it is exposed to current it's not meant to take.)

Some circuits work with split (+, –, and ground) power supplies, so make sure that you connect the leads of the pulser to the correct power points to avoid damage to the components.

Counting Up Those Megahertz

A frequency counter (or frequency meter) tests the frequency of a signal. You use a frequency counter to verify that a circuit is operating correctly. For example, suppose you create an infrared transmitter and the light from the transmitter is supposed to pulse at 40,000 cycles per second (40 kHz). With a frequency counter connected to the circuit, you can verify that the circuit is indeed producing pulses at 40 kHz, not 32 kHz, 110 kHz, or some other Hz.

You can use most models, such as the one in Figure 16-2, on digital, analog, and most RF circuits (radio transmitters and receivers are typical RF [radio frequency] circuits). For most hobby work, you need only a basic frequency counter; a $100 or $150 model should do just fine. And, some of the newer multimeters also have a rudimentary frequency counting feature.

Figure 16-2:
A digital frequency counter adds up frequencies.

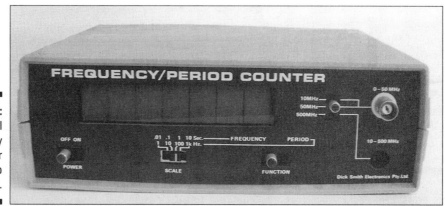

In a digital circuit, signals are limited to a range of zero up to about 12 volts. Voltages can vary widely in an analog circuit. Most frequency counters are designed to work with analog voltages ranging from a few hundred millivolts to 12 or more volts. Check the manual that came with your frequency counter for specifics.

Frequency counters display the frequency signal from 0 (zero) Hertz (cycles per second), to a maximum limit that is based on the design of the counter. This limit usually goes well into the megahertz; it's not uncommon to find an upper limit of 25 to 50 MHz. Higher-priced frequency counter models come with a prescaler or offer one as an option. A *prescaler* is a device that extends the useful operating frequency of the frequency counter to much higher limits. Go for the prescaler feature if you're working with high frequency radio gear or computers.

A Power Supply with a Changeable Personality

You use a power supply to replace batteries while building and testing circuits at your workbench. A variable power supply provides a well-regulated voltage output, generally ranging from 0 to 20 volts. The model in Figure 16-3 offers a variable output range of about 2 to 20 volts, as well as preset outputs of –5, +5, and +12 volts.

Figure 16-3: A variable power supply.

In addition to the voltage output of a power supply, pay attention to the current capacity. The higher the current rating of the supply, the more stuff it can power. Avoid a power supply with only a modest current output — say, less than one amp. You can't adequately drive all circuits with lower currents. Consider a power supply that delivers a minimum of two amps at +5 volts and at least one amp at any other voltage.

Making All Kinds of Signals

A function generator creates nearly pure signal waveforms for testing and calibration purposes. These gizmos are handy when you need to provide a known signal from one circuit to another circuit you're working on. For example, it's not uncommon to build circuits in stages, one piece at a time. Maybe you're building a little transmitter and receiver that work using just light, rather than radio frequency waves. You start with the receiver portion. The function generator can temporarily serve as your transmitter. When you get the receiver done, you can build the transmitter, knowing, thanks to the function generator, that the receiver is working properly.

Most function generators develop three kinds of waveforms: sine, triangle, and square. You can adjust the frequency of the waveforms from a low of 1 Hz to a high of between 20 and 50 kHz.

You need a frequency counter to accurately time a waveform. Some function generators come with a frequency counter built-in. If you have a stand-alone frequency counter, you can use it to fine-tune the output of the function generator.

Calling All Alien Worlds

A sweep generator is a type of function generator, but with a cool twist. A *sweep generator* produces signals that are somewhat different from the ones that a standard generator puts out, in that it sweeps the frequencies up and down. Not only does this sweep sound like E.T. calling home (connect a speaker to the output of the sweep generator to hear this effect), but it also helps you find frequency-sensitive problems in your circuits.

So what is this frequency-sensitive thing? Frequency-sensitive means that a circuit is sensitive to specific frequencies. Because of that characteristic, a circuit may function perfectly well at one frequency, but not at another. This would be bad for something like a radio receiver, which needs to operate over a range of frequencies. If you produce a range by sweeping the frequencies, you can more quickly see if your circuit is operating under all the conditions that you want it to.

A sweep generator varies the frequency of the output waveform, typically within pre-selected limits, such as 100 Hz to 1 kHz or 1 kHz to 20 kHz. You most often use sweep generators in troubleshooting audio and video equipment, where altering the frequency reveals bad components.

Some function generators also have a sweep feature, covering two functions with one tool.

Analyze This

A logic analyzer is like a souped-up oscilloscope (you can read about oscilloscopes in Chapter 10). It shows you the waveforms of several inputs or outputs of a digital circuit at the same time. You most often use a logic analyzer to test digital gadgets, and those folks well versed in the black arts of electronics find this analyzer useful.

One way you can use a logic analyzer is to check clock and data signals for a microcontroller. These devices require very specific timing relationships for various signals that you feed into them simultaneously. The logic analyzer lets you freeze-frame all the signals. Then you can see if a signal is missing or doesn't sync up with the others, as it should.

If you think that you're ready to try a logic analyzer, you can buy a stand-alone model or one that connects to your PC. Stand-alone units cost a pretty penny, and they're very sophisticated. Consider getting a less expensive logic analyzer adapter for your PC. These adapters connect to the USB, serial, or parallel ports of your computer. You need special software that comes with the adapter. Most PC-based logic analyzers handle between 8 and 16 digital inputs at one time.

A Trio of Testing Toys

Here are three testing tools that are somewhat specialized, but if you know a bit about them, you can impress people in electronics discussion forums. Oh, and you may just need one or more of them in a project someday!

Here are the three tools:

- ✔ **Spectrum analyzer:** This tool lets you actually see radio waves. Well, to be precise, you don't see the waves, but you see the radio energy created by them. The energy appears as a "spike" on an oscilloscope-like display. People sometimes use spectrum analyzers in amateur radio work to determine if a transmitter is on the fritz.

- ✔ **Signal injector:** This one literally injects a signal into an analog circuit. You use one of these puppies to test whether radios and televisions are in working order. You listen for the signal using a signal tracer or meter. You use the signal injector and tracer like you'd use the continuity test you perform with a multimeter, but this test goes further. To the trained ear (yes, these gadgets take some skill to use properly), you can tell just by the tone if components in the circuit may be bad.

- ✔ **Static meter:** If you have read much of this book at all, you know that static electricity can cause all kinds of problems for electronics components. You can use a static meter to scope out dangerous levels of static electricity on or near your workbench. If you get high readings, you can take steps to minimize the static. Remember that sensitive electronic components and static don't mix! Be sure to check out Chapter 2 for additional tips on reducing static electricity.

Where to Get Testing Tool Deals

I won't kid you — electronics test equipment can cost you a lot of money. Much of what you pay for is the accuracy of the device. Manufacturers of this or that doodad strive for high accuracy to tout their product in the market or meet necessary government regulations. If you're an electronics hobbyist working at home, you don't really need a very high level of accuracy. Usually you can get by with less expensive models.

When you buy any test gear, especially the special-purpose stuff mentioned in this chapter, don't automatically go for the high-priced spread. A pricey doohickey can't make you a better electronics tech. The low-end model is likely to be good enough for most hobby applications, and assuming that you take good care of it, it should last many years.

You also don't need to buy everything brand new. Used and surplus items can save you a ton of cash on electronics test equipment. Buying used and surplus has one disadvantage — most of this stuff doesn't come with instruction manuals. Sometimes you can buy the manual separately, or you may be able to find it online. Owners of popular test gear often scan the pages of their old equipment manuals and post them online for the benefit of others.

Check out these sources for used equipment:

- ✔ **eBay and other online auction sites:** Before you bid, check out other auctions, including those that have already ended, to see what the going price is for similar products. Set your bid accordingly, and use a proxy bidding feature so that you don't have to stay glued to your computer to stay on top of the bidding.

- ✔ **Electronics mail order and local surplus outlets:** These outlets are another good source for used test equipment and are handy if you don't want to wait for an auction to finish or you prefer to know the price up front.

Whether you use an auction, mail order, or a local store, be sure that the test gear you buy actually works. Have the seller guarantee that the equipment is in working order by giving you a warranty. You may pay a little bit more for it, but if you don't make sure that it works, and you're not so good at fixing broken test gear, you may just be buying an expensive paper weight. If you're brand new to electronics, have a more experienced friend or work associate check out the gear for you.

Pass up sellers, especially on eBay or other auction sites, who aren't willing to guarantee that their products are in working order. Plenty of sellers take the time to check out their wares and guarantee that the item won't be dead on arrival.

Chapter 17

Ten Great Electronics Parts Sources

*L*ooking for some great sources for your electronic parts? This chapter gives you some perennial favorites, both inside and outside of North America. This list is by no means exhaustive; you can find literally thousands of specialty outlets for new and used electronics. But the sources we list here are among the more established in the field, and all have Web pages for online ordering (some also offer a print catalog).

North America

Check out these online resources if you're shopping within the United States or Canada. Most of these outlets will ship worldwide, so if you live in a different country you can still consider buying from these stores. Just remember that shipping costs may be higher, and you may have to pay an import duty, depending on your country's regulations.

All Electronics

www.allcorp.com

All Electronics runs a pair of retail stores in the Los Angeles area and sends mail orders worldwide. Most of their stock is new surplus, meaning the merchandise is brand new but has been overstocked by the company. All Electronics has a printed catalog; the latest updates are available on their Web site. Be sure to check out the Web Only deals.

Allied Electronics

www.alliedelec.com

Allied Electronics is what's known as a stocking distributor. They offer goods from a variety of manufacturers, and they have most parts available for immediate shipping. A minimum order requirement may apply. Allied is geared toward the electronics professional, but they welcome hobbyists, too. The Allied catalog is *huge*. It's not practical to browse it cover to cover, but they do have a useful search feature.

B.G. Micro

www.bgmicro.com

Selling primarily surplus odds-and-ends, B.G. Micro has great prices and terrific customer service. You can buy either from their printed catalog or online. Check their Web site for the latest deals. Their stock tends to come and go quickly, so if you see something you especially like, be sure to order it now! Otherwise, some other eagle-eyed evil scientist may beat you to it.

Digikey

www.digikey.com

If you want it, Digikey probably has it. Like Allied Electronics, Digikey is a stocking distributor, carrying thousands upon thousands of items. Their online ordering system is particularly easy to use and includes price, available stock levels, and even quantity discounts. The site offers a handy search engine so you can quickly locate what you're looking for. Digikey will also send a free printed catalog, but to read the tiny print, you have to get out your glasses. The text has to be teensy-weensy to fit everything in.

Electronic Goldmine

www.goldmine-elec.com

Electronic Goldmine sells new and surplus parts, from the lowly resistor to exotic lasers. They've divided up their Web page by category, which makes ordering very easy. Most parts include a color picture and a short description. Be sure to check out their nice selection of project kits.

Fry's Electronics

www.frys.com
www.outpost.com

Fry's is a retail store chain with store locations primarily in Texas and on the west coast. The www.frys.com Web site is for the Fry's chain. Each store is overflowing with electronics, including ICs and resistors. The company's www.outpost.com Outpost Web site provides many (but not all) of the same products via mail order. The Outpost site is handy if you don't happen to have a Fry's store nearby.

Jameco Electronics

www.jameco.com

Jameco sells components, kits, tools, and more. They offer both convenient online and catalog ordering. You can browse the Web site by category, or, if you know the part number you're interested in — such as a 2N2222 transistor — you can find it by entering the part number into a search box. You can also use the search feature for categories of parts, such as motors, batteries, or project enclosures. Just enter the category term, and off you go.

Mouser Electronics

www.mouser.com

Similar to Allied and Digikey, Mouser is a stocking distributor with tens of thousands of parts on hand. You can order from their online store or their humongous print catalog. If you can't find it at Mouser, it probably doesn't exist. You can ask Mouser for a printed catalog and they'll send it to you. It's the same content as they have on the Web site, but we find it easier to browse for parts when they're printed on paper. Call us old-fashioned!

RadioShack

www.radioshack.com

RadioShack is perhaps the world's most recognized source for hobby electronics. They support thousands of stores worldwide and now ship many of the offerings in their extensive product line by direct mail. RadioShack still sells lots of resistors and capacitors in their neighborhood stores. But you may want to let your mouse do some online shopping when you need the more esoteric stuff, such as less-common integrated circuits or logic probes.

Outside North America

Electronics is popular all over the globe! Here are some handy-dandy Web sites you can visit if you live in places such Australia or the UK. As with North American online retailers, most of these folks also ship worldwide. Check their ordering pages for details.

Dick Smith Electronics (Australia)

www.dse.com.au

Electronics from Down Under. Dick Smith Electronics offers convenient mail order (the company ships worldwide) and has local retail stores in Australia and New Zealand.

Farnell (UK)

www.farnell.com

Based in the UK but supporting shoppers from countries worldwide, Farnell stocks some 250,000 products. You can order through their Web site.

Maplin (UK)

www.maplin.co.uk

Maplin provides convenient online ordering for shoppers in the UK, Western Europe, and other international locations. The company also supports dozens of retail stores throughout the UK.

Advice for Shopping Mail Order

For the most part, you can safely build up your cache of electronics goodies shopping by mail. Still, you may run into some hucksters and thieves out there, and it pays to be a little cautious. Here are a few do's and don'ts to keep in mind when conducting business by mail.

Do

When shopping by mail, be sure to:

- ✔ Understand exactly what you're buying, when the company plans to deliver it, and how much you're paying *before* you send any money.

- ✔ Carefully examine your credit card monthly statements for improper charges.

- ✔ Favor those companies that provide a mailing address and a working phone number for voice contact (not just fax). Sellers without one or the other aren't necessarily crooks, but lack of contact information just makes it harder to get hold of someone if you run into a problem.

- ✔ Be wary of companies that advertise by sending unsolicited spam e-mails. Also, be sure that the company Web site has an acceptable privacy policy regarding sharing your contact information.

- ✔ Verify shipping and handling charges and service fees before finalizing your order. These costs can add to the price significantly, especially for small orders.

- ✔ Check out the company before sending them a significant order (what qualifies as significant is up to you; it may be anything over $500, or it may be anything over $35). Check for a poor rating with the Better Business Bureau (or a similar institution for those readers outside the United States) in the company's hometown, in the appropriate news-groups, or in online chat rooms or bulletin boards.

- ✔ Add insurance, especially if you're ordering overseas. As a rule, once a package leaves the shipper, it "belongs" to you. If the shipment goes astray, you're left holding the bag. If you don't get insurance, you could be out money. Many shippers, such as UPS, automatically insure for up to $100. If your order is worth more than that, be sure to buy extra insurance.

- ✔ Determine added costs for duty, taxes, and shipping when buying internationally.

Don't

Okay, so now you know what *to* do. Here's what you should *avoid* whenever possible:

- ✔ Don't buy from a source unless you feel very comfortable about sending money to them.

- ✔ Don't give your credit card number over e-mail or on a Web page order form unless you know you're using a secure communications link. When you're using a site with security features, a little key lock appears on the status bar of your browser. This shows you that the communications between you and the site are encrypted with a code that thieves have a hard time breaking.

✔ Don't use a credit card to pay for goods from a company that you haven't yet dealt with if you can just as easily send a check or money order. This way, you limit the exposure of your credit card accounts to possible Internet fraud.

✔ Don't send money to foreign companies unless you're positive they're safe bets. While you're checking them out, be sure that they ship to your country.

New or Surplus?

Surplus is a loaded word. To some, it means junk that just fills up the garage, like musty canvas tents, or funky fold-up shovels that the U.S. Army used back in the 1950s. To the true electronics buff, surplus has a totally different meaning: Affordable components that help stretch the electronics-building dollar.

Surplus just means that the original maker or buyer of the goods doesn't need it any more. It's simply excess stock for resale. In the case of electronics, surplus seldom means used, as it might for other surplus components, such as motors or mechanical devices that have been reconditioned. Except for hard-to-find components — such as older amateur radio gear — surplus electronics are typically brand new, and someone still actively manufactures much of this equipment. In this case, surplus simply means extra.

The main benefit of shopping at the surplus electronics retailer is cost: Even new components are generally lower priced than at the general electronics retailers. On the downside, you may have a selection limited to whatever components the store was able to purchase. Don't expect to find every value and size of resistor or capacitor, for example.

Remember that when you buy surplus there is no manufacturer's warranty. Sometimes that lack of warranty is because the manufacturer is no longer in business. Though most surplus sellers accept returns if an item is defective (unless it says something different in their catalog), you should always consider surplus stuff as-is, with no warranty implied or intended (and all that other lawyer talk).

Chapter 18

Ten Electronics Formulas You Should Know

*F*ormulas take the guesswork out of electronics. Instead of dumping a bunch of components on the table and plugging them in any which way, the seasoned electronics experimenter builds new circuits with the help of a handful of formulas. These formulas help you determine specific values for things like voltage when you are designing electronic circuits.

You use these same formulas when modifying existing circuits. For example, you can apply the basic Ohm's Law formula for direct current (which you can find in the section titled "Calculating Relationships with Ohm's Law") to select a resistor so that a light emitting diode shines brighter or dimmer, as your prefer.

This chapter summarizes many of the more commonly-used electronic formulas that you encounter in your electronics work. The electronics world has used quite a few of these formulas for many, many years, but they still work just fine.

Calculating Relationships with Ohm's Law

Ohm's Law calculates the relationship between power, voltage, current, and resistance. Table 18-1 gives you the formulas you use to find these values.

Table 18-1	Formulas for Ohm's Law
Unknown Value	*Formula*
Voltage, in volts (V)	$V = IR$
Current, in amps (I)	$I = \dfrac{V}{R}$
Power, in watts (P)	$P = VI$
Resistance, in ohms (R)	$R = \dfrac{V}{I}$

Note that in Table 18-1:

V = voltage (in volts)

I = current (in amps)

P = power (in watts)

R = resistance (in ohms)

Check out this example: To find the power in a circuit consuming 100 volts at ten amps, multiply volts by amps (100 x 10 = 1000). So, you get the answer of 1,000 watts. You might use this figure to judge how big a fuse you can add to your circuit without damaging it, or how big an electric bill you're going to have at the end of the month.

Here's another example: To calculate the resistor that you need to handle a given amount of current through an LED, you use Ohm's Law like this:

$$R = \frac{V}{I}$$

Figure 18-1 shows a circuit made up of an LED, a resistor, and a battery (or other power source). You use this formula to calculate the value of R, the resistor.

Here's what the alphabet soup of V, I, and R means:

✔ **V (also sometimes noted as E):** The voltage through the LED. Because the voltage reduces when it goes through a diode, you have to subtract this voltage (about 1.2 volts for the typical LED) from the supply voltage. For example, V = 3.8 volts if the supply voltage is 5 volts and the drop through the LED is 1.2 volts.

✔ **I:** The current, in amps, that you want flowing through the LED. 20 mA is a reasonably safe value for almost any LED; a lower value makes for a dim light, and a higher value — much over 40 or 50 mA — may destroy the LED. Because you need to express I in amps, 20 mA becomes a fractional number: 0.020 amps.

✔ **R:** The resistance needed, in ohms, to limit the current to the LED.

Figure 18-1:
Use Ohm's
Law to
calculate
the value of
the current-
limiting
resistor that
you need for
an LED.

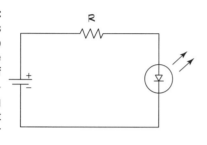

To continue the example for V and I we'll plug in some real numbers in place of the V, I, and R (which you can also see called out in Figure 18-1):

$$190 \text{ ohms} = \frac{3.8 \text{ volts}}{0.020 \text{ amps}}$$

See Chapter 1 and the Ohm's Law table in the yellow Cheat Sheet in the front of this book for more about using Ohm's Law.

Calculating Resistance

You can calculate the resistance of a single resistor in a circuit simply enough. But resistance changes when you add resistors in parallel or in series. For resistors in series, you add the resistance values together. For resistors in parallel, the result is a little less obvious.

Why bother with calculating resistance of multiple components? There are several good reasons:

✔ You can find resistors in only a limited number of common values. Some circuits call for a specific value that you can create only by inserting two or more resistors in series or in parallel.

✔ Resistors aren't the only components that exhibit resistance. For example, the windings of a motor also have a certain resistance. For some special applications, you need to calculate the combined effect of having these various resistances in a single circuit.

Calculating resistors in series

The formula for calculating resistors in series is pretty simple — just add up the resistances. Here's how it works:

$$Rt = R1 + R2 + R3 \ldots (\text{and any more, as needed})$$

In this case, R1, R2, R3, and so forth are the values of the resistors, and Rt is the total resistance.

For example, suppose you have two resistors rated at 1.2k ohms and 2.2k ohms. Add them together, and the resulting resistance is 3.4k ohms.

Calculating two resistors in parallel

Things are a little more complicated when you want to calculate two resistors in parallel. Here's the formula you use:

$$Rt = \frac{R1 \times R2}{R1 + R2}$$

R1 and R2 are the values of the two resistors and Rt is the total resistance. Given a 1.2k (1200 ohms) and a 2.2k (2200 ohms) resistor:

$$776.47 = \frac{2640000}{3400}$$

Now, to calculate three or more resistors in parallel:

$$Rt = \frac{1}{\frac{1}{R1} + \frac{1}{R2} + \frac{1}{R3}} \ldots (\text{and more as needed})$$

Here R1, R2, R3, and so forth are the values of the resistors. Rt is the total resistance.

Calculating Capacitance

You can use the formulas in this section to calculate total capacitance in a circuit. Note that the formulas are basically the inverse of the formulas for resistors, described earlier in this chapter. And, like resistors, the same logic applies for why you'd ever want to calculate capacitance of two or more capacitors together.

Calculating capacitors in parallel

To calculate the value of a string of capacitors in parallel, just add 'em up:

$$Ct = C1 + C2 + C3 \ldots$$

In this formula, C1, C2, C3, and so forth are the values of the capacitors; Ct is the total capacitance.

Calculating two capacitors in series

Use the following bit of math wizardry when you need to calculate the total capacitance of two capacitors wired up in series:

$$Ct = \frac{C1 \times C2}{C1 + C2}$$

In this formula, C1 and C2 are the values of the two capacitors; Ct is the total capacitance.

Calculating three or more capacitors in series

Got capacitors? Got lots of 'em? Well, if you're wiring them all up in series, you need to use a special formula to calculate the total capacitance:

$$Ct = \frac{1}{\frac{1}{C1} + \frac{1}{C2} + \frac{1}{C3}} \ldots$$

C1, C2, C3, and so forth are the values of the capacitors. Ct is the total capacitance.

Now, why would you ever want to add a bunch of capacitors together this way? One common reason is to provide a specific capacitance value for which there is no standard component. You sometimes need to do this with very sensitive circuits, such as radio receivers.

Calculating Units of Energy

The *watt-hour* is one of the most practical units of measure of energy; it's the ability of a device or circuit to do work. You calculate watt-hours by multiplying the power of the circuit, in watts, by the length of time you have the circuit on. The formula for calculating watt-hours is

$$Watt\text{-}hours = P \times T$$

In this formula, P stands for power, in watts, and T represents time in hours that it takes for power to dissipate. To calculate watt-seconds, also known as the *joule,* divide watt-hours by 3600.

Calculating RC Time Constants

Electronic circuits often use time constants to provide time delays or stretch the timing of signals. You most often construct them using a resistor and capacitor — hence the use of the term *RC.*

To complete the circuit, you connect the resistor and capacitor, as you can see in Figure 18-2, to some form of active component, such as an inverter or a transistor. You can select the values of the resistor and capacitor to produce a signal that lasts a specific amount of time.

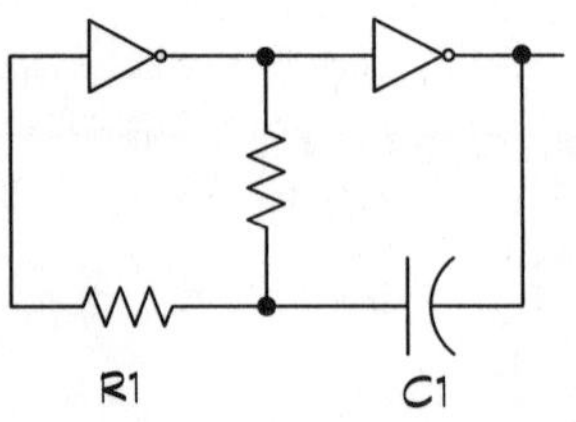

Figure 18-2:
A resistor and capacitor used to make a timing circuit.

R1   C1

RC circuits work because it takes a certain amount of time for a capacitor to discharge through a resistor. The larger the value of the resistor and/or capacitor, the longer it takes for the capacitor to discharge. Circuit designers use RC networks to produce simple timers and oscillators or to change the shape of signals.

So how do you calculate the time constant for a resistor-capacitor circuit? These circuits combine a resistor and a capacitor. Note that the capacitance value is in farads. Typical capacitor ranges are in microfarads and even smaller units, so the capacitance value is a fractional number.

$$T = RC$$

In this formula, T represents time (in seconds), R stands for resistance (in ohms), and C signifies capacitance (in farads).

For example, with a 2000-ohm resistor and a 0.1-uF capacitor, the time constant is 0.002 of a second, or two milliseconds. Table 18-2 shows some examples so that you can get the zeros right.

Table 18-2	Examples of Capacitance Value
Capacitor Value	*Capacitance Value for Calculation*
10 uF	0.00001
1 uF	0.000001
0.1 uF	0.0000001
0.01 uF	(0.00000001)

Calculating Frequency and Wavelength

The frequency of a signal is directly proportional to its wavelength, as the formulas in the following sections show you. You may find these formulas handy if you experiment with radio circuits (for example, when cutting a wire to a specific length to make an antenna). The following formulas express wavelength in meters and frequency in kilohertz.

Calculating frequency of a signal

Let's suppose you're interested in learning electronics so you can gab to folks all around the world on an amateur radio set. It would be useful for you to know all about radio frequencies. Radio frequencies, and the wavelength of the signals carried by those frequencies, work hand in hand. In amateur radio, you'll hear people say they're operating at such-and-such a wavelength. Here's how to calculate the frequency of that wavelength.

$$\text{frequency} = \frac{300,000}{\text{wavelength}}$$

Wavelength is stated in millimeters, not feet, inches, or a multiple of bunches of bananas. Frequency is stated in megahertz.

Calculating wavelength of a signal

You can use the same basic formula to calculate wavelength if you already know the frequency of the radio signal:

$$\text{wavelength} = \frac{300,000}{\text{frequency}}$$

The result is stated in millimeters. The frequency value is stated in megahertz. Here's an example. Suppose you're communicating with beings from another planet on 50 megahertz (50 million cycles per second). Plugging those numbers into the formula, you get:

$$6000 \, (\text{millimeter}) = \frac{300,000}{50}$$

Most folks talk about wavelength in meters, so there's one final bit of math to perform. As there are 1000 millimeters to a meter, the result is 6 meters. It seems you're talking to E.T. on the six-meter amateur radio band. Cool!

Appendix

Internet Resources

● ●

*I*n this appendix, we present a gaggle of interesting Internet sources for all things electronic. Businesses operate some of these sites, and individuals are at the helm of the others. We've taken the time to find what we consider the most useful resources to save you the time and bother.

Be aware that Web sites may come and go over time. If you try to visit a site and your Web browser can't find it, the site owner probably has moved on. That's life on the Internet! Try search engines, such as Google and Yahoo!, to find additional resources.

Figuring Things Out with Calculators

You can perform calculations on the sites in this section without having to look up equations or pick up a handheld calculator. Choose a Web site that covers the particular equation that you want to use:

- ✔ **Electronics Converters and Calculators** (`www.csgnetwork.com/electronicsconverters.html`): This site has calculators that perform Ohm's Law calculations, parallel resistance calculations, and resistor color code conversions, among other operations.

- ✔ **The Electronics Calculator Web Site** (`www.cvs1.uklinux.net/calculators/index.html`): Using tools that you find on this site, you can perform calculations for Ohm's Law, RC time constants, and a few other handy equations.

- ✔ **Bowden's Hobby Circuits** (`ourworld.compuserve.com/homepages/Bill_Bowden/homepage.htm`): The calculators on this site include the standard calculations for Ohm's Law, RC time constants, and resistor color-coded conversions. You can also find calculators for functions that you don't find on most other sites, such as a voltage divider calculator.

- ✔ **John Owen's Web Site** (`www.vwlowen.demon.co.uk/`): This site offers calculators for use with amateur radio and audio projects.

Gabbing about Electronics in Discussion Forums

Use the forums on the sites in this section to get answers to your questions about projects or general electronics. Every discussion area has its own style, so spend a little time on each site to decide which forum is right for you. Post your question and others who have lived through your quandary may provide the answer that you need.

We found the discussion groups on the following sites especially interesting and helpful:

- ✔ **All About Circuits Forum** (forum.allaboutcircuits.com): Here, you find both a general electronics discussion forum and a forum to ask for help from other forum members on any sticky projects.

- ✔ **Electronics Zone Discussion** (www.electronic-circuits-diagrams. com/forum/): This site has very active discussions on electronic circuits and projects.

- ✔ **EDAboard International Electronics Forum Center** (www.edaboard. com): Explore these active discussions about problems with projects and general electronics, along with several more specialized forums, such as one on PCB design.

- ✔ **Electronics Lab** (www.electronics-lab.com/forum/index.php): Here, you can find another good site with discussions on projects, circuits, and general electronics. Check out the Project Q&A section; here readers can post questions, and get answers, on the many projects provided in the projects area of the site.

Be sure to take the answers that you get on forums with a grain of salt. Think through the advice that you get before you build a project based solely on some well-meaning stranger's word.

Surfing for Robot Parts

Are you just enthralled by R2/D2? Does the robot in *Lost in Space* make your heart beat faster? If you're into robotics, you should like these Web sites that sell stuff for building small robots:

- ✔ **Acroname, Inc.** (www.acroname.com): This site belongs to a robot-specific online retailer. They offer R/C servos and other useful parts.

- ✔ **Budget Robotics** (www.budgetrobotics.com): Our very own author, Gordon McComb, just happens to run this site and it provides already modified servos, as well as small wheels and PVC plastic sheets for building 'bots.

- ✔ **Lynxmotion** (www.lynxmotion.com): You can find PVC or polycarbonate (Lexan) plastic sheets, Tamiya gearmotors and wheels, and R/C servos at this site.

- ✔ **The Robot Store** (www.robotstore.com): At this site, you can get R/C servos already modified for continuous rotation, Tamiya motors, and various kinds of wheels.

- ✔ **Solarbotics** (www.solarbotics.com): A reseller in Canada that sells small PVC sheets for making robots, this company also sells the kind of R/C servo motors and small wheels that you need to make a Smart Rover (which just happens to be one of the projects we walk you through in Chapter 15).

- ✔ **Tower Hobbies** (www.towerhobbies.com): One-stop shop for buying hobby stuff online. They sell a lot of R/C servos for making little robots, but those servos don't come already modified, so you have to do any modifications yourself (see the instructions for modifying servos for 'bots in Chapter 15). This site is also a good, reliable source for Tamiya motors and different styles of Tamiya wheels.

Many of these online retailers stock the kind of R/C servos that the Smart Rover project we describe in Chapter 15 uses.

Getting Up to Speed with Tutorials and General Information

The Web sites in this section all have worthwhile information. Browse through them to decide which sites meet your needs. We give the Kelsey Park School Electronics Club and the North Carolina State University Electronics Tutorial the highest marks, but all of these sites have cool and useful information:

- ✔ **All About Circuits** (www.allaboutcircuits.com/): This site contains a series of online books on electronics. They haven't yet posted some sections, but the material that they do have is well done.

- ✔ **Electronics Hobbyist** (amasci.com/amateur/elehob.html): Here, you can find interesting articles on various basic electronics topics.

- ✔ **Graham Knott's Web Site** (ourworld.compuserve.com/homepages/ g_knott/index1.htm): Enjoy exploring this site that Graham Knott, an electronics teacher at the University of Cambridge in England, has organized to make finding information on both beginning and intermediate electronics topics simple.

- **Kelsey Park School Electronics Club** (`www.kpsec.freeuk.com`): This site has a lot of good advice for newcomers to electronics projects, including a tutorial on how to read a circuit diagram, explanations of components, and a list of circuit symbols.

- **The North Carolina State University Electronics Tutorial** (`www.courses.ncsu.edu:8020/ece480/common/htdocs`): Contains good explanations of various electronics topics. Many of the illustrations are animated, which makes understanding the concepts easier for you.

- **Online Guide for Beginners in Electronics** (`library.thinkquest.org/16497/home/index.html`): Read brief introductions to several electronics topics here.

- **Williamson Labs Electronics Tutorial** (`www.williamson-labs.com/home.htm`): This site has some explanations of basic electronics concepts accompanied by good illustrations that you may have fun looking through.

Trolling for Printed Circuit Board Chemicals and Supplies

If you're into making your own printed circuit boards (see Chapter 12 for details), check out these Web sites for tools, chemicals, and supplies. Most of the sites in this section sell pretty much the same types of supplies, so we just list the Web pages without further description:

- **Circuit Specialists:** `www.web-tronics.com`
- **D&L Products:** `www.dalpro.net`
- **Ocean State Electronics:** `www.oselectronics.com`
- **Minute Man Electronics:** `www.minute-man.com`
- **Philmore-Datak:** `www.philmore-datak.com`
- **Press-n-Peel** (transfer film): `www.techniks.com`
- **Pulsar** (DynaArt transfer film): `www.pulsar.gs`

In addition to these sources, many general electronics resellers offer a limited selection of PCB-making supplies, as well.

Getting Things Surplus

Looking for some good deals? Try buying surplus. Because surplus merchandise comes and goes, you have to be on your toes to catch the good stuff — but if you're lucky, you can find great bargains. Try these online surplus electronics dealers:

- **Action Electronics** (www.action-electronics.com): This site sells both prime (brand new, direct from the manufacturer) and surplus items.

- **Alltronics** (www.alltronics.com): The inventory at this site is so huge that it may take you hours to get through it all. They have everything, from used motors to teeny-tiny electronics parts.

- **American Science & Surplus** (www.sciplus.com): A trusted and reliable reseller of everything surplus. They stock some small electronics parts, but go to these guys for the motors, switches, and larger stuff.

- **C&H Sales Company** (www.aaaim.com/CandH): A terrific source for older electronics parts, the kind you see in a 1950s sci-fi flick. They sell a ton of things, and if you live near their store in Pasadena, California, you can drop by and see everything in person.

- **Fair Radio Sales** (www.fairradio.com): This supplier has been around for about 50 years. They specialize in ham radio and military surplus, but they also have plenty of smaller bits and pieces to help you fill out your junk box.

- **Gateway Electronics** (www.gatewayelex.com): This site sells some kits and parts over the Internet. They also have a couple of stores — one in California and one in Missouri.

- **Marlin P Jones & Associates** (www.mpja.com): This site sells new and surplus electronics, test tools, and other goodies.

- **Skycraft Parts & Surplus** (www.skycraftsurplus.com): You can find a warehouse full of electronics and mechanical surplus, plus kits, test tools, and more, at this site.

In addition to these sources, be sure to also check out our top-ten list of online electronics outlets in Chapter 17.

Surfing for Circuits

Hungry for even more circuits to build? They're just a mouse click away! Thanks to the magic of the Internet you can find hundreds — no, make that thousands — of electronic circuits, from basic light and sound demonstrators to advanced projects for your car or boat. Here, then, are a few of the best:

- ✔ **Bowden's Hobby Circuits** (ourworld.compuserve.com/homepages/ Bill_Bowden/): This personal site from hobbyist Bill Bowden emphasizes the why, not just the how. Here you find both circuit descriptions and alternative design suggestions.

- ✔ **Discover Circuits** (www.discovercircuits.com): This member-supported site boasts over 8,000 schematics. Click on the List of Electronic Schematic Categories link to get to the several hundred categories.

- ✔ **Electronics Online** (www.electronicsinfoline.com): This ad-supported search engine for electronics projects and circuits offers hundreds of links that lead you to thousands of schematics in dozens of categories.

Glossary

· ·

*A*s with any field of study, electronics has its own lingo. Some terms deal with electricity and units of measure such as voltage. Other terms are tools you use in projects or electronics parts, such as transistors. Here are many of the terms you'll run into in your electronics life. Knowing these terms will help you become electronics fluent.

alkaline battery: A type of non-rechargeable battery. *See also* battery.

Allen screw or wrench: *See* Hex.

alternating current (AC): Current in which the direction of the flow of electrons cycles continuously from one direction to the other and back again. *See also* direct current (DC), Hertz.

amplitude: The amount of voltage in an electrical signal.

Anode: The positive terminal of a diode. *See also* cathode.

auto-ranging: A feature of some multimeters that automatically sets the test range.

AWG (American Wire Gauge): *See* wire gauge.

bandwidth: Relative to an oscilloscope, the highest frequency signal that you can reliably test, measured in megahertz (MHz).

battery: A power source that uses a process called electrochemical reaction to produce a positive voltage at one terminal and a negative voltage at the other terminal. This process involves placing two different types of metal in a certain type of chemical. *See also* alkaline battery, lithium battery, nickel-cadmium battery, nickel-metal hydride battery, zinc-carbon battery.

biasing: Applying a small voltage to the base of a transistor to turn the transistor partially on.

bipolar transistors: A common type of transistor. *See also* transistor.

breadboard: (also called prototyping boards or solderless breadboards) Plastic boards that come in a variety of shapes, styles, and sizes; they contain columns of holes that little slivers of metal connect electrically. You plug in components — resistors, capacitors, diodes, transistors, integrated circuits and so on — and then string wires to build a circuit. *See also* soldered breadboard.

buss: A common connection point.

cable: Groups of two or more wires protected by an outer layer of insulation, such as a common power cord.

capacitance: The ability to store electrons, measured in farads.

capacitor: A component that provides the property of capacitance (the ability to store electrons) in a circuit.

cathode: The negative terminal of a diode. *See also* anode.

circuit: A series of wires connecting components so that a current can flow through the components and back to the source.

cladding. A very thin sheet of copper that you glue over a plastic, epoxy, or phenolic base to make a printed circuit board.

closed circuit: A circuit where wires are connected, and current can flow. *See also* open circuit.

closed position: The position of a switch that allows current to flow. *See also* open position.

cold solder joints: A defective joint that occurs when solder doesn't properly flow around the metal parts.

commutator: A device used to change the direction of electric current in a motor or generator.

components: Parts used in electronics projects, such as a battery or transistor.

conductor: A substance through which electricity can move freely.

connector: Metal or plastic receptacles on a piece of equipment that cable ends fit into; an example of a connector would be a phone jack in your wall.

continuity: A test you perform with a multimeter to establish whether a circuit is intact between two points.

conventional current: The flow of a positive charge from positive to negative voltage; the reverse of real current. *See also* real current.

current: The flow of an electrical charge.

cycle: The portion of a waveform where the voltage goes from it's lowest point to the highest point and back again is one cycle. This cycle is repeated as long as the waveform is running.

DPDT: *See* double-pole, double-throw switch.

DPST: *See* double-pole, single-throw switch.

desolder pump: A device that sucks up excess solder with a vacuum.

diode: Components that limit the flow of current to one direction.

direct current (DC): Current in which the electrons move in only one direction, from the negative terminal through the wires to the positive terminal; the electric current generated by a battery is an example of direct current.

double-pole switches: A type of switch that has two input wires.

double-pole, double-throw switch (DPDT): A type of switch that has two wires coming into the switch and four wires leaving the switch.

double-pole, single-throw switch (DPST): A type of switch that has two wires coming into the switch and two wires leaving the switch.

electric current. *See* current.

electricity: The movement of electrons through a conductor.

electromagnet: Some form of coiled wire around a piece of metal (typically an iron bar). When you run current through the wire, the metal becomes magnetized. When you shut off the current, the metal loses that magnetic quality.

electromotive force: An attractive force between positive and negative charges, measured in volts.

electron: A negatively charged particle. *See also* proton.

embedded language interpreter: A program that runs inside the microcontroller that allows you to write your programs by using an easy-to-use programming language.

ESD (electrostatic discharge): *See* static electricity.

fillet: A raised area formed by solder.

flathead: A term used to describe both a screw with a flat head and single slot, and the screwdriver you use with it.

flux: A wax-like substance that helps molten solder flow around components and wire, and assures a good joint.

frequency: A measurement of how often an AC signal repeats (the symbol for frequency is f). *See also* Hertz.

gain: The amount that a signal is amplified (the voltage of the signal coming out divided by the voltage of the signal coming in).

gauge: *See* wire gauge.

ground: A connection in a circuit used as a reference (zero volts) for a circuit.

heat sink: A piece of metal that you attach securely to the component you want to protect. The sink draws off heat and helps prevent the heat from destroying the component.

helping hands clamp: (also sometimes called a third hand clamp) Adjustable clips that hold small parts while you're working on projects.

Hertz (Hz): The measurement of the number of cycles per second in alternating current. *See also* frequency.

Hex: (also called Allen) Both a screw with a squarish hole in the head and the wrench used with it.

high signal: In digital electronics, a signal at any value higher than zero (0) volts.

I: The symbol for current.

IC. *See* integrated circuit.

impedence: The measure of opposition in an electrical circuit to a flow of alternating current.

inductance: The ability to store energy in a magnetic field (measured in Henries).

inductors: Components that provide the property of inductance (the ability to store energy in a magnetic field) to a circuit.

infrared temperature sensors: A kind of temperature sensor that measures temperature electrically.

input/output ports. (also called I/O ports) Connections on a microcontroller through which signals are sent or received.

insulator. A substance through which electrons are unable to move freely.

integrated circuits (ICs): Components (often called a "chip") that contain several small components such as resistors or diodes.

inverter: A type of logic gate that has only one input. *See also* logic gate.

inverting mode: A process by which an op amp flips an input signal to produce the output signal.

jack. A type of connector. *See also* connector.

joule. A unit of energy.

lithium battery: A type of battery that generates higher voltage than other types, at about 3 volts. Lithium also has a higher capacity than alkaline batteries. *See also* battery.

live circuit: A circuit to which you've applied voltage.

logic gate: An integrated circuit that takes input values and determines what output value to use based on a set of rules.

low signal: In digital electronics, a signal at or near zero (0) volts.

microcontroller: A programmable circuit.

miter box: A tool used to make angled cuts with your hacksaw.

multimeter: An electronics testing device used to measure such things as voltage, resistance, and amperage.

negative temperature coefficient (NTC) thermistor: A resistor whose resistance decreases with a rise in temperature. *See also* resistor, thermistor.

nickel-cadmium battery (NiCad): The most popular type of rechargeable battery. *See also* battery.

nickel-metal hydride battery (Ni-MH): A type of rechargeable battery. *See also* battery.

n-type semiconductor: A semiconductor with contaminates added that causes it to have more electrons than a pure semiconductor.

ohm: A unit of resistance (the symbol for ohm is Ω). *See also* resistance.

Ohm's Law: An equation that allows you to calculate voltage, current, resistance, or power.

one-time programmable (OTP): OTP microcontrollers can only be programmed once.

open circuit: A circuit where a wire is disconnected, and no current can flow. *See also* closed circuit.

open position: The position of a switch that prevents current from flowing. *See also* closed position.

operational amplifier: (also called op amp) An integrated circuit used to boost an audio or other signal. An operational amplifier performs much better than an amplifier made from a single transistor. For example, an op amp can provide uniform amplification over a much wider range of frequencies than can a single-transistor amplifier.

oscilloscope: An electronic device that measures voltage, frequency, and various other parameters for waveforms.

oscillator: A circuit that generates waveforms. *See also* waveforms.

pad: Contact points on a printed circuit board used for connecting components.

phillips: A term used to refer to both a screw with a plus (+) shaped slot in the head and the screwdriver used with it.

photoresist: (also called sensitizer or resist) A light-sensitive chemical layer used in making circuit boards.

piezoelectric effect: The ability of certain crystals — quartz and topaz are examples — to expand or contract when you apply voltage to them.

pn junction: When regions containing boron and phosphorus are next to each other in a semiconductor, a pn junction is created.

positive temperature coefficient (PTC) thermistor: A device whose resistance increases with a rise in temperature. *See also* resistance, thermistor.

potentiometer: A variable resistor that allows for continual adjustment of resistance from virtually no ohms to some maximum value.

power: The measure of the amount of work that electric current does while running through an electrical component measured in Watts.

precision resistors: A type of resistor with low tolerance.

prescaler: A device that extends the useful operating frequency of a frequency counter.

proton: A positively charged particle. *See also* electron.

prototyping board: *See* breadboard.

p-type semiconductor: A semiconductor with contaminates added that cause it to have fewer electrons than a pure semiconductor.

pulse: A signal that alternates between high and low very rapidly.

pulse width modulation: A method of controlling the speed of a motor that turns voltage on and off in quick pulses. The longer the "on" intervals, the faster the motor goes.

R: The symbol for resistance.

RC time constant: A formula used to calculate the time it takes to fill a capacitor to two-thirds or discharge it to one-third of its capacity.

real current: The flow of electrons from a negative to a positive voltage.

relay: A device that acts like a switch in that it closes or opens a circuit depending on the voltage supplied to it.

resist: *See* photoresist.

resistance: The measurement of the ability of electrons to move through a material.

resistor: A component you add to a circuit to reduce the amount of electrons flowing through the circuit.

rosin flux remover: Available in a bottle or spray can, use this after soldering to clean any remaining flux to prevent it from oxidizing your circuit.

SPDT: *See* single-pole, double-throw switch.

schematic: A drawing showing how components in a circuit are connected together by wires.

semiconductor: A material, such as silicon, that has some of the properties of both conductors and insulators.

semiconductor temperature sensors: A kind of temperature sensor that varies the output voltage depending on temperature.

sensitizer: *See* photoresist.

sensors: Electronic components that sense a condition or effect such as heat or light.

series circuit: A circuit in which the current runs through each component sequentially.

short circuit: Where two wires are accidentally connected together and current goes through them rather than completing the circuit as intended.

sine wave: An output signal.

single-pole switch: A type of switch that has one input wire.

single-pole, double-throw switch (SPDT): A type of switch that has one wire coming into the switch and two wires leaving the switch.

60/40 rosin core: The ideal solder for working with electronics containing 60 percent tin and 40 percent lead (the exact ratio can vary a few percentage points) with a core of rosin flux.

slide switch: A type of switch you slide forward or backward to turn something (such as a flashlight) on or off.

solar cell: A type of semiconductor that generates a current when exposed to light.

soldered breadboard. A breadboard on which you have soldered components in place. *See also* breadboard.

soldering. The method you use in your electronics projects to assemble components on a circuit board to build a permanent electrical circuit; instead of using glue to hold things together, you use small globs of molten metal called *solder.*

soldering iron. *See* soldering pencil.

soldering pencil: A wand-like tool that consists of an insulating handle, a heating element, and a polished metal tip used to apply solder.

solderless breadboard: *See* breadboard.

solder sucker: A tool used for removing excess solder. The sucker is a spring-loaded vacuum.

solder wick: (also called solder braid) A device used to remove hard-to-reach solder. The solder wick is really a flat braid of copper. It works because the copper absorbs solder more easily than the tin plating of most components and printed circuit boards.

solid wire: A wire consisting of only a single strand.

spike: *See* voltage spike.

square wave. An output signal.

static electricity: A form of current that remains trapped in an insulating body.

strain relief: A device that clamps around a wire and prevents you from tugging the wire out of the enclosure.

stranded wire: Two or three small bundles of very fine wires, each wrapped in insulation.

stray capacitance: A condition where electric fields occur between wires or leads in a circuit that are placed too close together and electrons are stored unintentionally.

sweep generator: A device that produces signals that are somewhat different from the ones that a standard generator puts out in that it sweeps the frequencies up and down.

terminal: A piece of metal to which you hook up wires (as with a battery terminal).

thermistor: A resistor whose resistance value changes with changes in temperature.

thermocouple: A type of sensor that measures temperature electrically.

third hand clamp: (also called helping hands clamp) A small, weighted clamp that holds parts while you solder.

tinning: Heating up a soldering tool to full temperature and applying a small amount of solder to the tip to prevent solder from sticking to the tip.

tolerance: The allowed variation, expressed as a percentage, in the value of a component due to the manufacturing process.

traces: Wires on a circuit board that run between the pads to electrically connect the components together.

transistor: A device composed of semiconductor junctions that controls the flow of electric current.

V: The symbol for voltage; also commonly represented by E.

variable capacitor: A capacitor that consists of two or more metal plates separated by air. Turning the knob changes the capacitance of the device. *See also* capacitor.

variable coil: A coil of wire surrounding a movable metal slug. By moving the slug, you change the inductance of the coil.

variable resistor: See potentiometer.

voltage: An attractive force between positive and negative charges.

voltage divider: A circuit that uses voltage drops to produce voltage lower than the supply voltage at specific points in the circuit.

voltage drop: The resulting lowering of voltage when voltage pulls electrons through resistors (or any other component), and the resistor uses up some of the voltage.

voltage spike: A momentary rise in voltage.

watt hour: A unit of measure of energy; the ability of a device or circuit to do work.

waveform: Voltage fluctuations such as seen in a sine wave or square wave. *See also* oscilloscope, sine wave, square wave.

wire: A long strand of metal, usually made of copper, that you use in electronics projects. Electrons travel through the wire to conduct electricity.

wire gauge: A measurement of the diameter of a wire.

wire wrapping: A method for connecting components on circuit boards using wire.

zinc-carbon batteries: A low quality, non-rechargeable battery. *See also* battery.

Index

• **F** •